Preventing Headaches & Migraines

By Dr. M.N. Hossain
3rd Edition 2015

©Copyright by Dr M.N. Hossain, January 2015
All rights reserved. First printed in 2007.
ISBN: 978-1-312-94145-8

3rd **Edition.**
Published by Dr. M.N. Hossain.

No part of this book may be copied by any means without the written and signed permission of the author. This book is protected by US and International Copyright Laws.

ABOUT THE AUTHOR

Dr. M. N. Hossain is a practising GP in the South of England within the UK. He attained B.Med Sci. (Hons.) in 1986; M.B., B.S. in 1989 ,M.R.C.G.P. in 1994 and Fellowship of BMJ Case Reports in 2008.
A.F.H.E.A. in 2013 (University of Cambridge, UK).

ACKNOWLEDMENTS
Please see the list of references at the back of this book for details of the many written research studies & books , referred to during the compilation of this book. My fellow medical colleagues are also welcome to refer to this book as a reference.

Many thanks to my friends and colleagues for their anecdotes (stories) about the causes of their headaches & migraines.

DISCLAIMER
The anecdotes reported in this book are correct at the time of printing. New research studies will appear every year and ideas & theories about the causes of headaches and migraines will thus change in future. People reading this book are advised to consult their doctor before trying any of the ideas in it, especially the exercise sessions. They try these ideas at their own risk and no responsibility can be accepted by the author for any problems or complications arising from their personal actions.

CONTENTS

INTRODUCTION	i
CAUSES OF HEADACHES AND MIGRAINE	1
FOODS THAT CAUSE MIGRAINE	5
FOOD STORAGE, REFRIGERATION, PREPARATION COOKING AND RE-STORAGE	22
SODIUM CHLORIDE (SALT), POTASSIUM CHLORIDE (LOW-SALT) AND SODIUM BICARBONATE (BAKING SODA)	25
LIST OF FOODS	30
FOOD INDUCED MIGRAINE AND NARCOLEPSY	40
POTATO RESEARCH STUDIES	42
CHEMICALS IN OTHER FOODS	45
COELIAC DISEASE	47
PREVENTING POST EXERCISE MIGRAINE (PEM)	50
SUGGESTED EXERCISE SESSIONS TO TRY	65
MEDICATION FOR HEADACHES & MIGRAINES	67
EYE STRAIN	68
REFERENCES	69

INTRODUCTION

This book is based on detailed **health diaries, food diaries** & reading scientific research studies on headaches & migraines. It also reviews some of the literature on headaches & migraines. It integrates many different strands of thought & study into one concise logical rational book.

This book is a **summary** of over 17 years serious study & observations. We live in a clever advanced world full of well read, well informed clever people & most readers will easily understand the concepts in this book.

"A picture is worth a thousand words" because half the brain learns using images & pictures. The diagrams will help the reader to better understand the relevant anatomy & visualise the overlapping problems of **"Tension Type Headaches (TTH)"** and **"Migraine"**.

It is very important to be **well read & well informed** in life. I suggest that it is worth making a serious study of this book **if you suffer from migraines.** I suggest that the reader could read it 3 or 4 times over the period of a few months ,so that your brain memorises it, as repetition is an effective method of learning.

Migraine affects about 5 million people in the UK. It affects women much more than men and is often linked to the monthly menstrual cycle ("menstrual migraine"). **"Food**

PREVENTING HEADACHES AND MIGRAINES

induced migraine" is a well recognised disease entity. Often **two or more "trigger factors"** are needed to induce a full blown severe migraine of 100% intensity or severity e.g. strenuous exercise or housework may cause a delayed onset moderate migraine of 40% intensity 24 to 48 hours later, which is worsened by a migraine inducing foods eaten that day, and further worsened by "eye strain" at work or by prolonged driving.

New **"case reports"** indicate that many common foods such as tomato ketchup, salty foods, chemical food additives , colorants & preservatives, sodium bicarbonate (baking soda) or other "raising agents" are potent causes of headaches (listed in the sections below). Some of the patients were eventually found to have an **underlying bowel disease such as Coeliac Disease or Ulcerative Colitis**.

The information in the pages below will enable the reader to develop a **lifestyle strategy** (overall plan) to minimise his or her headaches & migraines.

These headaches have variously been described as sharp stabbing pains in the head; sometimes like a tight band around the head & sometimes as if the head is going to explode. They are quite distressing & make people suffer badly yet people look normal on the surface. In the past cynical people could not see any obvious physical deformity or disease & said people complaining of a headache or migraine were lying or just "faking it" to avoid work. We appreciate nowadays that is a real illness causing significant suffering & disability.

Dr.M. N. HOSSAIN

Sleep Deprivation
For some people who read this book , 8 hours sleep will be insufficient for good brain memory, concentration and brain function. Some people who read this book will need 9.5 to 10 hours sleep after 30 minutes of intense extreme exercise (for good memory). Many people seem not to understand this. People who sleep 10 hours at night often have a "photographic memory". Albert Einstein used to sleep 10 hours at night. Many famous sportsmen, ballet dancers and actors sleep 10 hours per night (often from 2am to 12 noon the next day). It is worth reading their **biographies** and articles.

Some symptoms of **sleep deprivation** are irritability & violence, annoyance, forgetfulness, poor concentration (causing accidents and poor performance), anxiety (stress), and daytime sleepiness with daytime **minisleeps and microsleeps.** It is especially bad if someone needs to sleep for 1 or 2 hours during the day as this often reflects bad sleep deprivation. If one has a late night on a Friday night & gets to bed at 1am, it is sensible to get up 10 hours later at 11am for good concentration & memory required for studying. **Winston Churchill** (the British prime Minister in World War II) worked long hours and is reported to have had mini sleeps in the daytime

The old fashioned method of children & teenagers sleeping from 9pm to 7am is sensible (9.5 to 10 hours) as hopefully this enables 3 complete REM (rapid eye movement) sleep-cycles for the brain. A good night's sleep works wonders.

PREVENTING HEADACHES AND MIGRAINES

People often find that after doing moderately heavy weight-training of 100 repetitions (reps.) with 25kg on a long barbell, they need 10 hours sleep for the next 2 days to recover. If they add 100 sit-ups to the exercise session he then needs an extra hour's sleep (about 11 hours sleep in each 24 hours, for the next 2 days). If they push themselves a lot & do 150 reps. of weight-training, they then need 12 hours sleep for the next 2 nights to recover. **The optimum exercise session for many people is thus 100 reps.** because it is difficult to find time for 2 hours extra sleep in a busy life.

Having discussed this with several university rugby players, footballers & cricketers one finds that such sleep requirements are common in sportsmen. It is worth remembering that most **professional athletes & sportsmen** sleep about **14 hours per day** because they get very tired from extreme exercising (training) twice daily (about 12 hours sleep at night and 2 hours sleep in the afternoons).

Saturday is a good day to do sports because people can sleep a lot on Sunday and eats lots of food and feel alright for work on Monday. If one has a full sports session on Sunday one might still feel tired on Monday morning and groggy.

Lack of sleep is considered a **trigger factor for headaches**. Sportsmen also eat **6 meals a day** rather than the usual 3 meals a day. A Diabetic sportsman will have 6 big meals a day and 6 Insulin injections a day, one with each meal.

CAUSES OF HEADACHES & MIGRAINE

If one knows the **cause** of headaches & migraines, one can avoid these & so suffer less. The full categorisation (classification) of the many different types of headaches by The **"International Headache Society"** is a whole book in itself & is outside the scope of this small book. There is an overlap between Tension Type Headaches (TTH) and Migraine.

Here are some common causes (triggers) of Tension Type Headaches (TTH) and Migraines:
1. Food substances (dietary triggers of migraine) e.g. tomato ketchup – **contains gluten**, salt (sodium chloride), lo-salt (potassium chloride), baking soda (sodium bicarbonate), cheese, artificial colourants, alcohol, choclate, old partly decomposed protein etc. See the detailed list below.
2. Exercise related headaches –
    (a) excessive exercise, physical activity or housework–post exercise headaches/ migraines from **exercise induced myositis**. Excessive pulling & stretching of the head/neck muscles & arteries; heavy lifting & straining (valsalva manoevre) or head-down position.
    Postcoital headache (after sexual activity).

PREVENTING HEADACHES AND MIGRAINES

    (b) under exercise & under activity.
3. Bright sunlight (causes eye ciliary muscle spasm), so wear sunglasses. (Also flickering artificial lights).
4. Menstrual related migraine in women.
5. Lack of sufficient sleep (insomina)
6. Dehydration
7. "Eye strain" from the muscles around the eyes focusing excessively for prolonged times.
8. Cervical Spondylosis (arthritis of the neck, degeneration of the neck vertebrae) in elderly people often causes neck & scalp pains.
9. High altitude (mountain climbing).
10. Crying (can cause a migraine afterwards)
11. Stress & anxiety can cause psychosomatic symptoms of headache.
12. Caffeine withdrawal (less caffeine intake at the weekends).
13. Overuse of pain killers containing codeine or caffeine (more than 30 per month) .(Medication overuse headache).
14. Some Medications cause side-effects of headaches or migraines. e.g. Calcium antagonists and Nitrates used to treat Hypertension.

CAUSES OF HEADACHES AND MIGRAINE

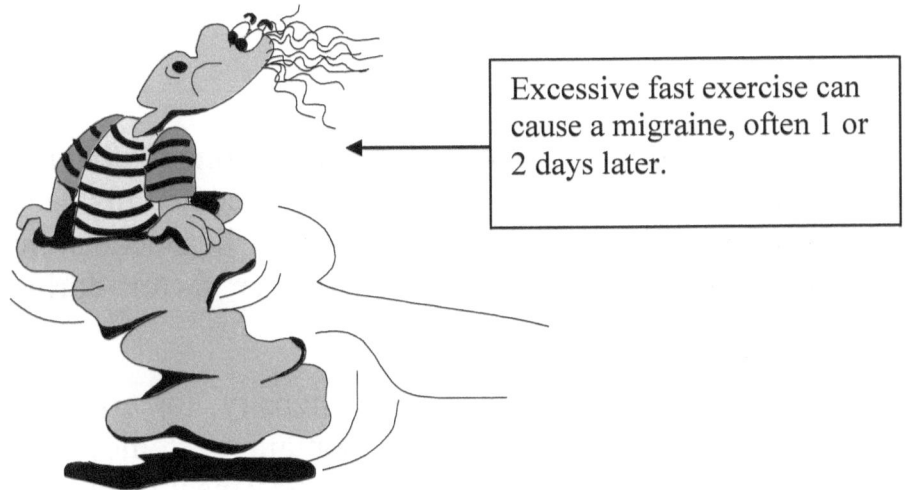

Excessive fast exercise can cause a migraine, often 1 or 2 days later.

Brain tumours, Hypertension (high blood pressure), anatomical abnormalities of the brain & spinal cord & brain diseases are uncommon or rare causes of headaches.

So most headaches are caused by the common causes (listed just above). **Lifestyle factors** are very important. It is important not to miss meals or lose sleep; to exercise regularly 2 to 3 times per week etc. Lack of food, water, sleep & exercise can all cause headaches.

One patient (subject) had a type of headache **"Cluster Headache"** which was very sensitive to loss of sleep & change in surrounding temperature. If the temperatures went under 70 F he got a bad sharp stabbing throbbing headache around the eye (probably from artery spasm). He slept for about 2 hours & was awake for 2 hours, alternately during every 24 hour period to control his bad headaches. His headaches were very **temperature sensitive** e.g. whenever he opened a fridge or freezer, or ate cold foods he got a

PREVENTING HEADACHES AND MIGRAINES

distressing headache. It is worth observing if your own headaches may also be "temperature sensitive".

Another subject noticed headache from **prolonged driving** for 2 hours. When he gently massaged & stretched his neck muscles the muscle-spasm was eased & the headache went away. **Eye strain** may also be a trigger factor during prolonged motorway driving.

He also got a mild migraine of 20% intensity, for up to 24 hours after a cup of **tea or coffee** (contain Caffeine). The caffeine caused forceful pounding of his heartbeat (inotropic effect of caffeine). Drowsiness was associated with the headache so he stopped drinking tea and coffee as this impaired his performance at work.

FOODS THAT CAUSE MIGRAINE

This list of foods is based on detailed and accurate health diaries & migraine diaries for over a year. **The foods we eat can make us quite ill.** There are written research studies listing chemicals in foods & food additives that are thought to cause migraine **(please see the list of references at the back of this book).** Most articles on migraine list chocolate, cheese, wine, alcohols and citrus fruits, but this book tries to go into more detail.

Some people react to **gluten in wheat, barley, rye and oats** and get a gluten induced headache or gluten induced migraine after eating such foods. They may have **gluten sensitivity or full blown Coeliac Disease.** So it is worth people keeping an accurate symptom diary and observing if they notice a headache within 1 to 12 hours of eating **gluten containing foods** (bread, pasta, tomato ketchup, salad creams, yoghurts etc.). It is worth these people reading books on Coeliac Disease and getting a **Gluten Free (G.F.) cookbook.**

Headache diaries indicate that food substances are the commonest contributor for headaches.

PREVENTING HEADACHES AND MIGRAINES

Be very careful what you eat!

Most books on **"Food intolerance" & adverse food reactions** advise a low dairy, low wheat, low nut, low histamine & low amine diet. This largely overlaps with the diet advice given below.(Reference – "The Allergy Bible"). **So it is a very good idea for migraine sufferers to also have a low wheat, low dairy' low nut' low histamine ,low salt & low amine diet (see references below). "Histamine poisoning" is a recognised entity.** We know that wheat protein, cows' milk protein & peanut protein anger the body's immune system. A **"low allergy"** diet is advisable.

Having a **low dairy diet** is straightforward e.g. avoid cows' milk, yoghurts & cheese. Skimmed milk is less concentrated. Try soya milk & soya yoghurts. Even a 70% reduction in dairy intake may benefit you. Sweetened soya milk is available (sweetened with apple juice) but **the citric acid seems to cause migraines !**

Having a **low gluten or Gluten Free (G.F.) diet** is difficult; corn flakes are not GF; rice is often stored in silos with

FOODS THAT CAUSE MIGRAINE

Barley and are not **"pure" rice** so rice pops are not GF ; pasta, noodles or bread made from rice flour , corn or soya flour; plain boiled rice is alright with meals instead of wheat etc.. If a person reacts to many foods it is best to eat foods that are labelled as **"natural" or "organic" (chemical free).** The Turkish, Italian or Asian grocery stores in London get their supplies from their own countries and their supplies may be more "pure" (uncontaminated) and "organic" than the supermarkets.

It is scientifically sensible to have a big breakfast, big lunch & small supper e.g. for breakfast one could have 1 bowl of cereal and an apple or a pear. Having most of your food intake during the day will prevent hunger cravings so you will be able to **avoid chocolates, sweet foods, crisps & snacks.** (These sweets and snacks often contain chemical food additives, excess salt, acids and alkalis that can cause migraines).

If you suffer from migraines you do need to keep your own health diary, food diary, exercise diary & migraine diary over the months & years, to record migraine "**trigger factors**" that affect you personally. Make a food list (diet list) of your top 20 or 30 foods that suit you & do not cause migraine. **Your own personal observations** of your headaches/ migraines & other health matters are very important. **AVOID THE FOODS LISTED BELOW IF YOU SEEM TO REACT TO THEM.**

Many common foods that we would usually consider okay, can actually cause bad migraines if a person has sensitive muscles & arteries.

PREVENTING HEADACHES AND MIGRAINES

A tiny amount of these foods may cause just a **mild** headache/ migraine of just 10% intensity lasting 1 to 12 hours. A larger amount may cause a **moderate** migraine of 30 % intensity lasting all day (1 to 12 hours). So a tiny dose of these foods is Ok but do not take an "**overdose**" of these foods. The severity of the migraine is "**dose related**".

If a person has very "**sensitive arteries**" his scalp & skull arteries go into spasm causing a pulsating, throbbing, pounding headache with throbs in timing with his heartbeat. This may be associated with nausea, vomiting (**abdominal migraine**) or blurred vision. This may be on one side or both sides of the skull. This can be caused or worsened by **chemicals or substances from foods** which are absorbed into the bloodstream within 7 hours of eating. The headache/migraine often lasts 1 to 12 hours and is maximal at about 7 hours after eating the offending foods.

The correct term for this is "**food intolerance**" and not "**food allergy**" because this is not directly arising from a reaction of the body's immune system (defence system).
Coeliac Disease is a true food allergy which involves a reaction of the immune system.

One **mince pie** (containing brandy or cider or other **alcohols)** may cause a mild migraine of 10% intensity but eating 2 mince pies at Christmas may cause a moderate migraine of 40% intensity lasting 12 hours, needing Paracetomol tablets to kill the pain. Beware alcohols in other sweets & desserts e.g. crème broulee.

FOODS THAT CAUSE MIGRAINE

The **wine industry** seems to use many **chemical additives** (e.g. acids, anti-oxidants, stabilisers, cleaning agents, fruit flavourings, oak powder, bentonite) yet these are apparently not mentioned on the label. The use of anti-freeze (ethinyl glycol) in 1985 at methanol in 1987 (in European wines) caused deaths & illness. This is mentioned in the book "The Great Wine Swindle" by Malcolm Gluck (2008). So it is not surprising that many people gets headaches/migraines after wines or other types of alcohol. (Reference: The Daily Mail Newspaper in the UK, Saturday 27th September 2008).

Many people get a mild **"Ice cream headache"** but if a person has sensitive arteries she may get a bad **"Ice cream migraine"** for 12 hours afterwards. One small ice cream may not cause a problem but if a person has two ice creams this may cause a bad 40% to 100% migraine (overdose). Many ice creams contain the **chemical food additives** E471 (mono and di-glycerides of fatty acids), **citric acid (flavouring) and artificial chemical colours (curcumin, annatto)**. The acids & alkalis seem to be a problem. I suspect it is these chemicals that cause the migraine. **Organic and gluten free** ice cream without such food additives is worth trying but seems difficult to find. Some

PREVENTING HEADACHES AND MIGRAINES

companies and manufacturers do not use such additives and one needs to search carefully for these.

One subject noticed that 2 packs of crisps is an "overdose" & can cause a migraine of 50% intensity. The likely offending food additives listed on these packets of crisps were: cheese flavour, **flavour enhancers (monosodium glutamate (MSG),** disodium 5-ribo nucleotide, flavouring, cheese powder, butter acids, colour (paprika extract, sulphite ammonia, caramel) , yeast, onion powder, pepper, colour (annatto), wheat. The MSG is well known for causing ill health. The **sodium content & salt content of foods is a big problem** because excess intake of oral sodium has profound bad effects on the heart & blood vessels (please see the section below). **Avoid salty foods such as crisps, chips, salted nuts etc.** Also there are strong naturally occurring chemicals in **poorly cooked potatoes** which can cause headaches and migraines.

He noticed that she got a mild 10% migraine for 1 to 4 hours after eating sweets containing chemical additives (E129, E104, E110, E102, E113 and E132). These **chemical food additives** were :

E129 = Allura red (food red dye), a chemical azo dye.
E104 = Quinoline yellow (a synthetic chemical coal tar dye).
E110 = Sunset yellow (a synthetic chemical coal tar dye & azo dye).
E102 = Tartrazine (also a synthetic chemical azo dye) which is known to cause migraine in children.
E132= Indigo carmine (another synthetic chemical coal tar Dye (blue colour)).
E133= Brilliant blue (synthetic coal tar dye) used to create

FOODS THAT CAUSE MIGRAINE

blue/green colour.
E210 = Benzoic acid (used as a preservative).
E222 = Sodium bisulphate (known to react with E210 listed above).
Some others are :
E100 = Curcurmin, an extract of turmeric. It is extracted using solvents such as methanol, hexane and acetone. Trace amounts of these solvents may cause migraine.
E160B = Annatto. This extracted using sodium hydroxide, potassium hydroxide, acetone, hexane & methanol. I suspect trace amounts of these chemicals cause migraine.
E163 = Anthocyanins. These are also extracted using solvents such as methanol & ethanol.

Many of these chemicals are known to cause allergic reactions, nettle rash & hyperactivity in children etc. It is advised that most of them should be avoided in hyperactive children. Some people are sensitive to such chemical food additives. Their side effects are all listed in **the book "E for additives"** (please see the reference section at the back of this book).

Mushy peas from a can contained the chemical food additives E102 and E133 (paints & colorants), causing a headache of 20% intensity for 6 hours afterwards.

One subject also noticed that 4 soya biscuits or 2 slices of soya or rice bread (made from rice flour or soya flour) is an "overdose" & can cause a migraine of 50% intensity because she had **sensitive arteries**. This may be due to food

PREVENTING HEADACHES AND MIGRAINES

additives or substances **(sodium bicarbonate (baking soda), yeast fungus & salt)** in the biscuits or bread.

He also noticed that eating one Danish pastry was OK but eating 2 pastries was an overdose & caused a 40% migraine for 12 hours **(possible reaction to gluten in wheat ,sodium bicarbonate (baking soda as a raising agent) or the yeast fungus in baked products).** The softest fluffiest cakes & pastry probably have the highest concentration of these chemicals & seem to be potent causes of headache.

One subject mistakenly used soya milk on her breakfast cereal that had "gone bad". There was a sediment of thick soya cream at the bottom of the carton (presumably with a concentration of particular chemicals). This caused a two stage migraine lasting a total of 7 hours consisting of :
- (a) a migraine 1-2 hours after eating (of 5% intensity), and then
- (b) a worse migraine 3-7 hours after eating (of 20% intensity).

He took two Paracetomol tablets to relieve this. He sometimes noticed a headache for 36 hours after eating a soya yoghurt or soya milk - but this eventually turned out to be because of **wheat (gluten) products**. The timing of the wheat induced migraine had overlapped with the migraine induced by these other foods.

The **chemical food additives** listed on the offending carton of soya milk were: Acidity regulators, Tri-Calcium phosphate, Tri-Potassium citrate, Di-Magnesium Phosphate. Other brands contained concentrated apple extract **(citric**

FOODS THAT CAUSE MIGRAINE

acid), maldodrextin, Vitamin E, Riboflavin, Vitamins D2 + B12. After eating, these chemicals are slowly absorbed into the bloodstream from the gut and affect the arteries in the body (including the skull causing migraine) for a seemingly long time. **Citric acid & citrates seem to be very potent at causing migraines. So beware oranges, pineapples, grapes, lemons and their juices. Avoid all acidic foods & fruits e.g. containing citric acid, acetic acid (vinegar). Onions contain oleic acid which makes our eyes sting.**

After a single **strawberry (containing histamine)** he noticed a mild 5% migraine for 30 minutes. If he ate a whole bowl of strawberries she got a migraine of 80% intensity for about 6 hours & needed to take Paraceteomol tablets to ease the pain.

He was unable to eat raw or poorly cooked tomatoes, **tomato puree, tomato ketchup containing gluten or tomato sauces (high in histamine content as there is often 176g of tomatoes concentrated per 100g of ketchup)** . He ended up avoiding all tomato products & pizzas. (Reference : see the book "The Allergy Bible" listed at the back of this book, page 200).

Eating a whole small bag of spicy chilli nuts or brightly coloured sweets caused a bad migraine all night, lasting 18 hours in total. He took 6 Paracetomol tablets during that time. Spicy salted nuts often contain the **chemical food additive MSG (monosodium glutamate)** which has a bad reputation. The MSG and **high salt content (sodium chloride or potassium chloride) causes transient hypernatraemia (high blood sodium level)** has bad effects on the heart & circulation.

PREVENTING HEADACHES AND MIGRAINES

Some foods cause quite **delayed** migraine e.g. after a Chinese takeaway, the migraine from **monosodium glutamate (MSG)** may appear 8 to 10 hours after the meal. If a subject had Indian snacks, onion bhaji, samosas which contain **wheat gluten** etc. at 7pm, the migraine may start at midnight & last all night. The headache diaries indicate after soya products & chocolate, the migraine starts within 1 hour of eating, is maximal from 7-10 hours after eating & can often last up to **36 hours, because the food stays in the bowel reservoir for this duration** releasing gluten and irritating the immune system for this time.

If you have a craving for chocolate, the supermarkets do have **gluten free chocolates in the "Free From" Section.** Like many of the foods listed below it can cause a delayed migraine lasting up to 12 hours (or even longer depending how much chocolate you have eaten). It is probably the amine and other types chemicals in chocolate that causes the headaches. The health diaries indicate that the **chemical food additives E500 (sodium carbonates), E501 (potassium carbonates) and E341 (calcium orthophosphates) and wheat flour (gluten)** can cause migraine - these are used in chocolates made be **some manufacturers/ companies** but not others. So identify the chocolate products made by particular companies, that seem okay for you.

Some meal combinations are especially bad at causing migraine because they contain 2 or more migraine inducing substances in foods e.g. **cheese sandwiches** (cheese & wheat bread); tuna sandwiches (tuna & wheat bread); spicy pizzas with peppers, tomatoes, halopinos. The new trend of

FOODS THAT CAUSE MIGRAINE

adding chillies & spices to snacks, nuts & crisps is undesirable and regrettable. The supermarkets are getting better at making ready made **"Free From" (Gluten Free)** frozen foods such as pizza, frozen meals and chicken steaks; and baked GF breads and cakes.

To understand "food induced migraines" you need to understand the cumulative effect of gluten reactions, food chemicals & substances in the blood stream from many foods, as they day progresses e.g.

One subject ate 2 chocolate biscuits at 12 noon (causing a 5% migraine) + 1 piece of pizza with tomato puree (causing a 10% migraine) + another chocolate biscuit at 4pm (causing a 5 % migraine) + a pack of crisps (causing a further 10% migraine) + eyestrain at work caused a further 30% migraine, so by 7pm he had a bad accumulated migraine of 60% intensity. He found **wheat products containing gluten** were especially bad probably because of **yeast fungus & sodium bicarbonate** present in these. **Salty foods** clearly affected his circulation & caused headaches afterwards. The wheat itself was not to blame in his case.

On another day he ate two slices of rice bread at breakfast & two more at lunch. He noticed a bad headache of 40% intensity lasting all day for 12 hours. He suffered for two days like this. This was not normal for him & this did not occur previously. He concluded that some substance in the "rice bread" (probably **yeast fungus or sodium bicarbonate** as a "raising agent") was causing these migraines. He took four Paracetomol tablets each day.

PREVENTING HEADACHES AND MIGRAINES

He realised that as a teenager he had no migraines so he reverted back to the diet he had then. He remembered with fondness the **nice school meals, boiled rice, boiled potatoes, freshly cooked meat** (without chemicals, food additives, colorants & preservatives). He changed back to his healthy home cooked childhood foods and had fewer migraines.

He also noticed a 2 stage migraine after eating 2 bags of children's chocolates which contained **gluten in wheat:**
- (a) an initial mild headache after an hour lasting 1-2 hours (Stage 1)
- (b) a delayed headache (of 80% severity) after this, lasting for 10 hours (Stage 2).

He also noticed that **chocolates** by some companies seemed OK for him. The problematic chocolates contained **wheat flour with gluten; the additives** E500 (sodium carbonates), E501 (potassium carbonates) and E341 (calcium orthophosphates). The sodium, potassium & calcium can affect the heart & circulation (as explained below).

When he had a bad respiratory infection (flu like illness), repeated **violent coughing and sneezing** caused a bad Tension-Type Headache (TTH) of 100% intensity by the end of the day, from increased tension in the scalp muscles and raised pressure inside the skull.

He had a chocolate gateaux for dessert after a nice lunch at a restaurant (containing **gluten** in wheat flour, **yeast fungus & sodium bicarbonate**). He noticed a bad migraine of 100% severity for 8 hours afterwards. This was distressing & he

FOODS THAT CAUSE MIGRAINE

took four Paracetomol tablets that day. He read in his health books that chocolate, cheese, wine & alcohols can all cause migraine. There was probably a lot of **gluten in wheat baking soda (sodium bicarbonate)** in this soft fluffy cake.

He was eventually found to have **Coeliac disease** (a reaction to wheat protein (gluten)) and migraine. Whenever he had **tiny traces of wheat gluten** in foods he got a bad migraine for many hours afterwards. He even needed to avoid cereal bars, sweets, desserts, Arabic sweets, and batter covering fish or chicken containing tiny traces of **wheat**. He found that many of the commercially manufactured wheat-free and gluten-free foods caused migraine, probably because of high sodium content (sodium bicarbonate (baking soda), other chemical additives; and yeast fungus).

He also had "**lactose intolerance**" & "**fructose intolerance**" because of bowel dysfunction (bowel disorder, bowel malfunction) caused by his **Coeliac disease**.

"**Lactose intolerance**" meant that his bowel was unable to break down lactose on cows' milk, so he got embarrassing flatulence (flatus, bowel wind) 1 to 10 hours after eating dairy products, cakes, cappuccino, cows milk etc. So he avoided such foods to reduce the socially embarrassing flatus.

"**Fructose intolerance**" meant that his dysfunctional bowel was unable to break down the plant sugar fructose (present in many sweet fruits). This a caused sugar fermentation in his bowel & embarrassing flatus 1 to 12 hours after eating sweet fruits e.g. figs, prunes, pears, pineapple, grapes,

PREVENTING HEADACHES AND MIGRAINES

melon, mangoes, lychees, apricots, plums etc. So he avoided these foods also. He could eat these in the evenings but then had noisy flatus at night. He realised that he got itchy bottom & blood stained mucous from the bottom, 3 to 7 days after eating **chillies, peppers & spices**. He stopped all foods containing chillies and peppers which solved the problem. He got **watery diarrhoea** an hour after he drank 2 glasses of pineapple juice because his diseased bowel could not digest the fructose.

His skin Eczema (dermatitis) also improved on a **low histamine diet**, low amine diet & low Nickel diet. Marmalade often contains the following food additives : oranges (contain sulphites), **added gluten**, gelling agent (pectin), citric acid, acidity regulator (sodium citrate), orange oils. I think it is probably the citric acid & citrate that causes a mild migraine (of 20% intensity) after eating marmalade, lasting all day. Some jars of marmalade, conserve & jam may have a high content of such chemicals wheras others do not.

He also noticed that after a tangy tasting Indian yoghurt dish which contained **gluten in wheat flour** his heart raced fast (tachycardia) & he got a bad throbbing pounding migraine in his head, in timing with his heartbeat all night. His heart was pounding forcefully because of the **immune system reaction to gluten** . He noticed a similar reaction after eating Indian cheese (paneer) and Indian sweets which were also made with **wheat flour containing gluten**.

Very spicy acidic curries contained histamine type chemicals in the food causing "**histamine poisoning**" (similar to the histamine release which occurs during an

FOODS THAT CAUSE MIGRAINE

allergic reaction) – please see the list of references at the back of this book.

It took him many years to realise that he got a migraine 2 to 10 hours after eating some types of old reheated poorly stored fish, (cod steaks, canned tuna, canned pilchards), chicken & omlette. He took 2 Paracetomol tablets each time. He was quite shocked when he finally realised this. This may have been a **mild reaction** to partly decomposed protein, but not the full blown allergic reaction. Freshly cooked fish eaten immediately seemed to be okay but fish or prawns that were **a day old and reheated** caused migraine (possibly higher content of amine type chemicals (histamine, tyramine) in old partly degraded foods. The high content of biologically active **amines and histamine content** made his heart pound forcefully & caused a throbbing pulsating headache (inotropic effect of histamine and amines). This is "histamine poisoning" – the food had literally poisoned him. The rule was to have meals that were **freshly cooked and quickly eaten (especially the protein part)**.

Fish fingers made by some companies contained chemical food additives & colorants (e.g. turmeric, capsanthin, curcurmin, yeast, **wheat with gluten**, annatto) and caused migraine. He made a note of the good companies & brands which used no such food additives & only bought those special brands. So it is important to write down and buy only **one's favourite brands of foods.**

Some brands of frozen battered squid contain many chemical food additives e.g. salt, raising agents (disodium diphosphate, sodium bicarbonate), garlic flavour, lemon

PREVENTING HEADACHES AND MIGRAINES

flavour, colour (riboflavin) and **wheat flour with gluten**. These caused a migraine lasting about 18 hours in one sufferer.

Cartons of rice pudding caused a mild 10% migraine (possible adverse reaction to bacterial or fungal proteins in this). A cheese omlette caused a 20% migraine for 1 hour after eating (probable high monoamine, tyramine & histamine content naturally occurring in cheeses). A mushroom omlette is preferable. He finally realised that he sometimes had **giddiness, vertigo & nausea** as symptoms of a migraine, especially if he leant his head forwards. This was probably a malfunction of the brain's balance system because of altered blood supply within the skull from the migraine **(basilar migraine)**. Abdominal symptoms (nausea, vomiting, abdominal pains) are recognised as being related to a migraine **(abdominal migraine)**. Some people can even get weakness of one side of the body **(hemiplegic migraine)**.

He was very surprised that he noticed migraines and wheezing after eating biolive yoghurts containing live bacteria. This is surprising because there are not chemical food additives in such yoghurts but there is often **added gluten. Gluten is a glue like substance which is added to may foods** by food manufacturers – see any **Gluten Free Cookbook.**

It took him many years to realise that he got mild bleeding from his gums for a few hours after eating peanuts. He was mildly allergic to peanuts too! He sometimes got rectal bleeding for a few days after eating **peanuts, chillies, peppers & spices.** Very mild spicy dishes were okay e.g.

FOODS THAT CAUSE MIGRAINE

Korma curries in Indian restaurants. When cooking, he put some mild **non irritant, non acidic, herbs/ spices** into dishes e.g. 1 clove of garlic, half a teaspoon of coriander powder and half a teaspoon of cumin powder, 3 or 4 tablespoons of olive oil and half a cup of water.

Most people get a **"hot flush"** after eating chillies, from the naturally occurring histamine in such foods. This is not healthy and is like having an **allergic reaction (histamine poisoning)**. People also get headaches/migraines & skin irritation after such foods – please see the list of references at the back of this book.

Even big rugby players have collapsed after trying **very hot spicy curries** such as Vindaloo ot Madras curry in 1989 and 2013. These spicy foods affected their blood pressure and they collapsed. One rugby player got a burst stomach ulcer after such foods.

One subject also noticed strange smelling urine & burning on passing urine (dysuria) after eating citrus fruits, orange juice & certain brands of spicy garlic sausage (presumably acidic/ alkaline chemicals or food additives in these). After high protein meals (meat & fish) the urine was a strong yellow colour (from excretion of protein degradation products).

PREVENTING HEADACHES AND MIGRAINES

FOOD STORAGE, REFRIGERATION, PREPARATION, COOKING AND RE-STORAGE

Very detailed migraine symptom diaries indicate that one needs to be meticulous about food storage, refrigeration, cooking & re-storage to prevent migraines. It seems that **old protein** (e.g. **meat, fish, omlette**) , even just 4 hours old can undergo some kind of change (decomposition, degeneration, degradation) that causes a headache or migraine. There may also be **growth of bacteria or fungi on the old stored foods** which cause an adverse reaction, resulting in a migraine. There are some interesting written research studies on this topic (**please see list of references at the end of this book**) which mention the growth of Penicillium fungi (moulds) on sausages in Argentinia; the bacterium Lactobacillus Plantarum on local fermented meat products in Thailand; putresceine production in decomposing foods; photo-degradation of foods; the study or Microwave Assisted Wet Degeneration (MAWD) on seafoods; MDA (malon-dialdehyde) produced from decomposing meat & fish proteins; the browning (chemical oxidation) of foods; and the degeneration of frozen fish & meat protein (amino acids).

One subject tried to save money by using less gas & cooking 2 portions of fish each time he cooked. He ate one immediately & placed the other portion in the fridge once it had cooled. The reheated food caused a migraine the next day, wheras the **freshly cooked fish** did not. It is a real challenge to cook fresh, healthy balanced meals (with carbohydrate, protein and vegetables) twice daily when one has a busy life & hectic schedule. When **protein** is broken down it produces amine type chemicals. Old rice &

FOODS THAT CAUSE MIGRAINE

vegetables stored overnight in a fridge (once cooled) seemed to be OK.

Omlette, fish or chicken burger cooked at home at 7am before work, caused a migraine when eaten at 12 noon (just 5 hours later), even if it was quickly placed in a fridge at 4 degrees Celsius, immediately or arrival at the workplace. It is best to **cook meat or fish from the frozen state** (rather than defrosted). Most home freezers are set at –20 Celsius & fridges are set at +4 Celsius.

On days off work is it better to cook at 12 noon because the food will cool in about 3 hours & can be placed in the fridge overnight. If one cooks at 7pm, the food has not cooled sufficiently by bedtime to be refrigerated overnight, and may be mistakenly left out overnight at room temperature causing some **partial decomposition**.

Unfortunately many takeaway places do not sell healthy boiled rice or boiled potatoes. So he sometimes ended up having **freshly cooked burger & unsalted chips, without tomato ketchup or salad cream and only with green lettuce & cucumber as salad.** He threw the soft bread bun (wheat containing gluten) away each day (containing sodium bicarbonate (baking soda) & yeast fungus) . Cooked rice & vegetables taken to work & placed in the fridge were okay. It was the partly degraded **protein** that caused migraine. There was a microwave at work & he was able to heat up food. **Cans of organic peas, sweetcorn & vegetables** could be quickly reheated in the microwave at work.. Apples & pears were easy to carry to work as "portions" of healthy fruit & vegetable fibre. Cans of corned beef seemed to cause migraine **(contained wheat; partly decomposed protein and high salt content).** So it is probably best to avoid all

PREVENTING HEADACHES AND MIGRAINES

canned meat & fish products as the protein is likely to be partly decomposed..

The method of cooking is also important. **Char-grilled or barbecued foods** may have traces of coal tar (soot) containing aromatic hydrocarbons (e.g. toluene, benzene) which may possible cause migraine. It is important not to burn the food as the black soot (burn) marks may contain similar chemicals. This needs people to follow the cooking instructions meticulously and thoroughly, without burning it and to have a good understanding of English (if one lives in the UK) .

It is preferable that the fresh meat or fish in a takeaway is **cooked on a metal hotplate or fried in oil,** rather than being char grilled with burnt soot (coal tar) marks on the food. You can often observe the food being cooked in front of you in a **takeaway or fast-food shop**. Spicy chicken from takeaways seems to cause migraines after it is eaten (contains wheat coating with **gluten** chemical food additives, peppers & spices in the coating or batter). Most takeaway shops do not seem to have an oven to use for cooking .

So the above observations have widespread implications as to the shops from which migraine sufferers source their foods, the cooking methods & food storage

FOODS THAT CAUSE MIGRAINE

SODIUM CHLORIDE (SALT), POTASSIUM CHLORIDE (LOW-SALT) AND SODIUM BICARBONATE (BAKING SODA) IN FOODS

It is unfortunate that these chemicals are commonly added to many foods and **have profound & bad effects on the heart & circulation** – please see the list of references at the back of this book which lists the relevant research studies in the past 10 years.

Sodium Chloride (salt)

It is best to **avoid salty foods** e.g. crisps, salted chips & peanuts. Have a low salt diet. Avoid extra salt on foods e.g. have unsalted fish & chips. The usual healthy low cholesterol, low fat, low sugar, high carbohydrate (CHO), high vegetable diet is also advised.

The inside of the body is like a chemical soup & an imbalance of some chemicals causes malfunction (disorder, dysfunction). **Excess oral salt intake from foods causes high blood sodium levels (Hypernatraemia** which may be temporary or long term). Salt overdose causes **toxic effects**. There are many written research studies confirming that high levels of sodium from "salt" has serious adverse affects on the heart & blood vessels. These include high blood pressure (Hypertension); hypertrophy (growth) of heart muscle & kidney tissue (Left Ventricular Hypertrophy (LVH) and renal hypertrophy) causing a heavier heart & kidneys; higher systolic blood pressure, arterial pressure and altered blood flow pattern through the heart valves; salt induced cardiac ventricular fibrosis (scar tissue in the heart

PREVENTING HEADACHES AND MIGRAINES

from collagen deposition) after a heart attack, also causing remodelling of the aortic valve in the heart; higher blood pressure during sleep in young men ; thicker heart & blood vessel walls via stimulation of vessel reactivity & growth, from increased transmembrane sodium gradient of vascular muscle ; impaired left ventricle (LV) relaxation and filling due to increased afterload of the heart; increased levels of nitrous oxide (NO) causing further heart fibrosis ; higher extracellular sodium concentration; faster blood flow ; dilatation & reduced flexibility of the arteries ; **cardiovascular ageing & reduced life expectancy;** the Renin-Angiotensin-Aldosterone System (RAAS) is adversely affected such that Aldosterone (a mineralocorticoid steroid hormone) function is altered causing Hypertension, endothelial dysfunction, heart & blood vessel remodelling ; salt-sensitive hypertension via Aldosterone (explained above); **increased heart & kidney sympathetic nerve activity with physiological release of noradrenaline (norepinephrine)** from altered endogenous Angiotensin 2 hormone activity in rats affecting heart baroreflex regulation by Angiotensin 2; release of ANF (atrial natriuretic factor) from the heart which acts via the GC-A (Gyanylyl cyclase-A) receptor in the kidney; increased cyclosporine level causing thicker heart muscle in rats (mammals very similar to humans); altered hydroxyproline concentration causing reduced left ventricle relaxation; salt induced hypertension in rats ;increased forearm vascular resistance (FVR) ; the presence of dysfunctional responses to exercise in hypertensive patients who have enlarged heart ventricles, obesity & high salt intake.

The release of **noradrenaline causes vessel constriction (arterial spasm)** in the body , including the skull & could

FOODS THAT CAUSE MIGRAINE

explain the observation of a migraine for up to 18 hours after eating salty foods, maximal at about 7 hours after such foods are eaten, resulting from **transient hypernatraemia** (high blood salt levels). The severity of the headache/migraine probably depends on the amount of salt eaten.

Potassium Chloride (low-salt)

Potassium chloride is added to some foods (e.g. crisps) as a supposed healthier alternative to common salt (sodium chloride) but still has bad effects on the heart & blood vessels. Its' use seems misguided – please see the research studies listed at the back of this book. Excessive oral Potassium chloride intake will cause high blood potassium levels (transient **hyperkalaemia**).

Potassium depolarises heart muscle causing **noradrenaline release from the sympathetic nerve terminals of the heart,** by stimulating the alpha receptors in the heart (increased adrenaline levels by stimulation of PKC ecto-5 nucleotidase) . Oral Potassium is used to treat heart dysrhythmias (long QT syndrome Type 2). Potassium chloride **induces vessel constriction (vasospasm)** via increased calcium influx in vessel smooth muscle cells. Research studies show a 43% increase in LDF (laser doppler **flow) in brain blood vessels** & increased tissue oxygen levels. Potassium chloride induces **depolarisations of brain (cerebral) blood vessels** in mammals. Potassium chloride & phenylephrine & endothelial chemicals stimulate secretion of ANF from atrial cells. Potassium chloride also affects calcium levels inside & outside heart atrial cells.

PREVENTING HEADACHES AND MIGRAINES

The release of noradrenaline & vasospasm effects of the brain blood vessels may explain the headache/ migraine noticed by some people after eating certain crisps & other foods containing Potassium chloride. The timing is similar to that of salt (see section above). The release of noradrenaline probably explains the **forceful pounding throbbing headache in timing with the heartbeat (inotropic effect of noradrenaline).**

Sodium bicarbonate (baking soda)
Baking soda is an **alkali** is widely used as a raising agent in bread, rolls, baps, pizzas , sandwiches , cakes, biscuits etc. It too has profound bad effects on the heart & vessels. It is probably the high **sodium** content from **sodium** bicarbonate that affects the circulation causing a headache/ migraine 1 to 12 hours later, maximal about 7 hours after eating.

Strong **sodium** bicarbonate is used as an effective treatment for hypotension (low blood pressure) & dysrhythmias & blood acidosis. This alkali affects the acidity/ alkalinity in heart cells & thus **affects the strength & rhythm of the heart**. Also, the high sodium intake causes **transient hypernatraemia** (high blood sodium level), resulting in release of **noradrenaline** from the heart & kidneys thus probably causing a pulsating throbbing forceful headache/ migraine , in timing with the heartbeat (inotropic effect of noradrenaline).

So it is logical for migraine sufferers to **avoid foods** containing sodium chloride, potassium chloride & baking soda (as explained in the sections above). **Some manufacturers** use lots of baking soda (sodium

FOODS THAT CAUSE MIGRAINE

bicarbonate) in their bread & wheat products & one needs to keep an accurate health diary of these. Presumably these manufacturers use less actual wheat & make more profit by doing this. The softest fluffiest breads & cakes usually have the highest concentration of raising agents (baking soda, yeast fungus) so it is best to avoid these.

The **calcium content** of foods is relevant also. As mentioned above calcium affects the smooth muscle heart & circulation too.

PREVENTING HEADACHES AND MIGRAINES

LIST OF FOODS

Research studies list the following foods & additives as migraine inducing **(listed in logical alphabetical sequence)** : alcohols, apples, aspartamine, bananas, benzoates, beta phenyl ethyl amine (in chocolate), caffeine, canned fish e.g. tuna, pilchards (contain histamine), cheese, Chinese takeaways (often contain MSG), chocolate (often contains **gluten**), citrus fruits, colouring & flavouring agents, cured meats, cola drinks, coffee, dairy products, food additives (sodium nitrate, monosodium glutamate (MSG), hot dogs, meat, milk, minerals, nitrates (food additives), nuts, onions (oleic acid), phenolic flavinoids, pizza,, phenylalanine, pork, tartrazine, tea, tomatoes (high natural **histamine content**), tyramine (in cheese), vitamins, wheat products. Also some acidic **histamine containing** vegetables (peppers, chillies), raw tomatoes & raw strawberries (contain histamine) are listed as causing migraines.

If a person reacts to many foods then he should be checked for an **underlying bowel disease such as Coeliac Disease (C.D.) or Inflammatory Bowel Disease (I.B.D.).** Observing the stools (bowel motions(is important as he may notice whole undigested foods (peas, carrots, sweetcorn, peppers, chillies, tomatoes) in the stools indicating a bowel disease or disorder. This indicates a bowel **malabsortion syndrome.**

FOODS THAT CAUSE MIGRAINE

There are also written research studies confirming that many foods have an immunological cross-reaction with latex rubber & other substances, thereby worsening skin eczema (dermatitis). I have included these studies in the **list of references** given at the end of this book. The foods listed for eczema are almost the same as those listed as causing migraines. Even some doctors are poorly informed about these research studies & seem to disbelieve that foods affect the skin.

We live in an advanced, highly organized society where food production & processing occurs on a huge scale. There are many vested financial interests for many companies & individuals, in food production. It is already well known that that chemicals & food additives are added to many foods. This is well documented in other books written over 20 years ago e.g. the book "E for Additives" listed in the references below.

These foods are listed below in logical alphabetical sequence. This list is obviously not fully comprehensive or perfect but will give a good idea of what foods to avoid to reduce migraine frequency & severity.

A
Alcohols (fermented plant alcohols in desserts, sweets & fruits). Avoid old desserts that have gone bad & smell of alcohols.

B
Barbecue (BBQ) foods with charcoal (soot) deposits – contain coal tar chemicals that can cause migraines.

PREVENTING HEADACHES AND MIGRAINES

Biscuits-(beware of food additives, **wheat gluten**, baking soda (**sodium** bicarbonate causing **sodium overload**) and yeast fungus in some makes of biscuits). See section on "bread" below.

Bread – certain types of bread by certain companies may contain food additives e.g. **bleach** to make white bread look white. Traces of **pesticides, fungicides, insecticides** from the wheat/grain storage silos may also be present. The **wheat gluten itself, sodium bicarbonate (causing sodium overload) or yeast fungi** may cause an adverse reaction causing a migraine.

 - Rice bread – made from rice flour available from some supermarkets contains food additives e.g. yeast, vegetable glycerine, stabilisers, emulsifiers.

- Burgers – beware **wheat with gluten**; food additives in processed meat burgers & pies. The bread bun may contain baking soda & yeast fungus.

Bananas – Phenol type natural chemicals in bananas can cause drowsiness & sleepiness.

<u>C</u>
Cakes-from some shops or supermarkets may contain food additives. Contain gluten in wheat flour.

Café meals - (food additives, wheat with gluten or chemical additives in chicken nuggets & chips, fish & chips, fried chicken)

Charcoal coated foods (from barbecues)

Cheese - cheese slices, cheesecake, Indian cheese (Muthar paneer) contains wheat flour . Buy for **"organic" or "natural" cheeses**.

Chips, spiced "fries" or potato wedges.

FOODS THAT CAUSE MIGRAINE

Chillies are high in **natural histamine chemicals**.
Chilli powder - (often contains a bright red dye (Sudan III)).
Chinese takeaways contain **(food additives e.g monosodium glutamate (MSG))**, fried rice, soy sauce. Proper Oriental restaurant food using good quality ingredients seems OK but costs more money.
Chocolate - the tiniest piece of a chocolate chip can cause a severe migraine. (v.potent) SO AVOID ALL CHOCOLATE PRODUCTS. (chocolate chip cookies, chocolate ice cream, chocolate desserts, Chocolates, chocolate sauce, hot chocolate drinks). Contain **gluten in wheat flour**. Possible reaction to chocolate protein too.
Christmas cakes & puddings - (contain alcohols e.g. brandy, cider)
CITRUS FRUITS & JUICES (Citric acid) is v.potent.
Coffee- (caffeine)
Conserves - (jams, marmalades, processed fruits); may have **gluten added.**
Crème Broulle - (alcohol)
Crisps - some crisps contain **wheat flour** and food additives e.g. cheese powder, cheese flavourings Paprika, sulphite ammonia caramel, onion powder, Peppers, colour (annatto), disodium 5 ribonucleotide. Eating two packs of crisps may be too much for a migraine sufferer. The **high salt content & sodium content** is a problem. Avoid salty foods.
Curries - (chillies, peppers, spices). So only eat very mild Korma curries. Red chilli powder contains a red dye (Sudan III).
Cakes - often contain **wheat with gluten**, food additives, **sodium bicarbonate (baking soda) causing sodium overload,** yeast fungus or chocolate. The softest fluffiest cakes are the worst because they have the highest content of

PREVENTING HEADACHES AND MIGRAINES

raising agents (sodium bicarbonate & yeast fungus).

D
Danish pastries – see section on "bread" above.
Doughnuts (food additives) – see section on "bread" above.

E
Eggs - in many countries chickens are fed maize with yellow colourants added to make the egg yolk look more yellow(E161, xanthophylls). Try **"organic" eggs** which are supposed to be fed on organic chemical-free foods.

F
Fig roll biscuits (food additives, **wheat with gluten**) –see section on "bread" and "cakes" above.
Fish & prawns – beware old partly decomposed protein.
Flapjack – contains **gluten in oats**.
Fudge - a tiny amount causes a severe migraine (v. potent). May contain gluten in wheat & chemicals.
Fish- processed frozen fish pies & fish steaks from some shops or supermarkets may contain wheat, food additives, colourants. Beware old fish (partly decomposed protein).
Fruit tart – contains food additives & wheat gluten
Fruit juices (frozen & defrosted) – beware citric acid.

G
Garlic bread (contains sodium bicarbonate (baking soda) and yeast fungus & **gluten in wheat**) – see the section on "bread" above.

H-
Haldi (turmeric) often contains a bright yellow dye (paint).

FOODS THAT CAUSE MIGRAINE

Hash browns – contains potato, food additives and possibly wheat gluten.

Halopinos (peppers, chillies) – contain histamine. They have a sharp acidic irritant taste & smell.

Humous (beware some shops may have humous with **added gluten and food additives).**

I

Indian snacks, onion bhaji & samosas (from supermarkets)- contain **wheat gluten**.

Indian sweetmeats – often very brightly coloured (artificial colours and wheat gluten).

Icing on cakes – may contain wheat.

Ice cream "ice cream headache" or "ice cream migraine". Contains food additives, flavourings, colorants. V. potent.

J

Jam (colorants, food additives, "processed" fruits); may contain **added gluten**.

K

Kebabs (be aware that these are often processed meats containing chemical food additives & charcoal soot deposits from barbecueing), chillies, peppers, spices, partly decomposed protein, high salt & sodium content. Delayed migraine 1 to 12 to hours later.

L – None yet

M

Marmalade (citric acid, sodium citrate, food additives); **added gluten**.

PREVENTING HEADACHES AND MIGRAINES

Meat - beware processed meats which often contain added chemical preservatives & other food additives.

Muthar Paneer (Indian cheese)- contains **wheat gluten**.

Malted bread (malt loaf) often contains food additives; **gluten in barley, malt or rye..**

Mince pies – Christmas pies contain alcohol e.g. brandy, cider. The pastry will contain **sodium bicarbonate & yeast fungus and gluten wheat.**

<u>N</u>

Nuts (Beware dry roasted nuts are often coated in a chemical powder containing Monosodium Glutamate (MSG), chilli powder, spice extract & paprika). Avoid spicy chilli nuts. A tiny amount causes a severe migraine for up to 18 hours (v. potent because of **MSG & high salt content causing sodium overload**).

<u>O</u>

Onions - if raw or poorly cooked in sandwiches & curries (contain oleic acid which makes our eyes water).

Onion bhaji (Indian snacks) – contain **wheat gluten**.

OLD FOODS – avoid foods that are old or out of date as they are partly decomposed.

<u>P</u>

Pasta sauces (beware acidic sharp tasting, sharp smelling sauces with peppers, containing natural histamine).

Pies (processed meats with **wheat; added gluten and food additives).**

Pizzas (these are often "processed" & frozen, containing tomato products, **wheat gluten**, high content of baking soda (sodium bicarbonate), yeast fungus & food additives). Get **GF pizzas** from supermarkets.

FOODS THAT CAUSE MIGRAINE

Peppers (if raw, uncooked or barely cooked have histamine chemicals). They have a sharp acidic tangy smell & taste.
Mushy peas –contain g chemical colorants E102 and E133.
Potatoes – avoid **poorly cooked potatoes.**

Q – None yet

R
Rice bread - from supermarkets (food additives or substances) ; **baking soda** may cause headaches.
Rice cakes (plain and flavoured); not pure ? May have been stored with **a silo with barley (which contains gluten)**; spicy flavoured ones contain food additives e.g. cheese powder, chilli powder.

Red onions – in sandwiches.
Spiced rice e.g. Pilao rice from supermarkets may have food colorants. Restaurant food is usually OK but costs more money.

S
Salad creams – some makes often contain food additives. Beware tiny amounts of some types of coleslaw or salad cream in burgers from high street Burger shops; may contain **added gluten. Use "organic" or "natural"** salad creams.
Salmon- salmon farms often feed the fish with added colorants, to make the flesh appear more pink (see the book "E for Additives" in the reference section below.
Samosas (Indian snacks) – contain **wheat gluten**.
Soups – contain food additives. Beware acidic Minestrone soup (has peppers & has a sharp tangy acidic smell & taste).
Soy sauce – food additives; may contain **added gluten**.

PREVENTING HEADACHES AND MIGRAINES

Strawberries – if raw, uncooked or poorly cooked (high in natural histamine). Strawberry jam is well cooked & is OK.
Sweets – food colorants used for brightly coloured sweets.
    -may contain **added wheat or gluten**.
Soya milk, soya bread & soya yoghurt. (food additives).

T
Tea (caffeine)
Tomatoes – are high in natural histamine chemicals if raw, uncooked or poorly cooked.
Tomato puree – avoid this. Is a potent cause of migraine. Red colourants in tomato puree; may have **added gluten**.
AVOID TOMATO KETCHUP – has added gluten. Get "organic" or "natural" ketchup or special GF ketchup in the "Free From" section.

Tuna – is naturally high in histamine. Avoid jacket potatoes with tuna, Coleslaw.
Turkish or Greek meals – beware chillies, peppers, spices, tomatoes. If the food has a sharp acidic taste it is high in histamine or acids.
Turmeric (haldi) powder often contains a bright yellow dye.
BEWARE BRIGHTLY COLOURED SPICES & FOODS.

U – None yet

V
Vegetable burgers – contain food additives e.g. vegetarian cheese powder, cheese flavourings, onion powder, peppers (paprika), mature cheese, citric acid; wheat gluten.
Vinegar (this is acetic acid). Beware vinegar added routinely to salads, salad creams, & kebabs in shops, takeaways &

FOODS THAT CAUSE MIGRAINE

restaurants.

<u>W</u> – <u>Wheat products</u> – avoid bread, sandwiches, rolls, buns, baps, pizzas, cakes, biscuits etc. see section on "bread" above. Strangely, pasta may be OK with a bland pasta sauce because this usually contains no **baking soda or yeast fungus (raising agents). Contain gluten.**

<u>X</u> – None yet

<u>Y</u>
Yoghurts – anthocyanins or colorants in some yoghurts. Tangy tasting yoghurts may also contain histamine. Biolive youghurts may contain **added gluten**. Childrens' yoghurts with **no added artificial colourants or additives** seem to be okay. **Use "organic" or "natural" yoghurts.**

<u>Z – None yet</u>

PREVENTING HEADACHES AND MIGRAINES

FOOD INDUCED MIGRAINE AND NARCOLEPSY

The written research studies on **Potatoes** indicate that contain very strong natural chemicals. They used to be called **"devil's food"** and can cause headaches and other **food intolerance reactions**. Green potatoes are especially high in these chemicals. It is important that all potatoes are well cooked so that these chemicals are broken down. The reported symptoms are **severe drowsiness (narcolepsy)** within an hour of eating the potatoes; then followed by a **migraine** for up to 12 hours. This is a **potato induced narcolepsy**.

This is probably only a problem if a person has an underlying bowel disease such as **Coeliac Disease** because the **diseased bowel** lacks the ability to properly digest the potatoes. People with a normal bowel possess the normal enzymes and ability to correctly break down foods.

Chemical food additives would not be broken down in such ill people because of the **diseased bowel,** thereby going into the blood stream and affecting the brain and vessels.

Simply frying potatoes as chips for 15 minutes or making crisps is insufficient to destroy these strong chemicals. The only good method seems to be to **part boil them until soft; then to roast them in the oven for 90 minutes.** The **duration of cooking** is important. Roasting them for 30 minutes is insufficient and can still cause a mild headache or migraine.

FOODS THAT CAUSE MIGRAINE

It is best to avoid foods containing **potato flour** so read the labels and go for foods made from **rice flour or corn flour.**

Spicy samosas have a **high histamine content** and in such susceptible people this histamine goes into the blood stream and can affect the brain causing severe sleepiness in the daytime (narcolepsy). So these people should avoid spicy foods in the daytime. This can be called a **histamine induced narcolepsy** from **histamine intolerance.**

Bananas contain naturally occurring **phenol type chemicals** and reports suggest that these chemicals can cause severe brain drowsiness (narcolepsy) also. This can be called a **banana induced narcolepsy and headaches.** Such **ill people** should avoid bananas.

Most peoples are aware that **red kidney beans and green beans contain strong naturally occurring chemicals** and need to be cooked for at least 30 minutes to destroy these. Such vegetables should **not be eaten raw** as the strong chemicals can cause abdominal cramps and vomiting. It is important that food in general, is well cooked – slow boiled for about 45 minutes.

So it would be interesting to check people who suffer from **Narcolepsy** for such an underlying bowel disease. A **suitable research study** can be planned where patients are given varying doses of these foods to eat and the effects are measured at 30 minute intervals for 12 hours afterwards.

PREVENTING HEADACHES AND MIGRAINES

POTATO RESEARCH STUDIES
The **written research studies on potatoes** (Solanum tuberosum) report that people can get **allergic reactions** to raw potatoes such as runny eyes and nose (rhinoconjunctivitis), contact urticaria, angioedema, Oral Allergy Syndrome (OAS), wheezing, collapse (systemic anaphylaxis), positive skin tests (PST), positive RAST, and eventual Eczema. There are immediate (Type 1) and delayed (Type 3) immune system reactions to potatoes. **(Please see the list of references at the end of this book).**

Peeling raw white potato can cause asthma attacks, runny nose and eyes in some **atopic (allergic) people**.

The **potato protein (Patatin)** cross reacts with tomato and latex protein and other fruit proteins. So latex allergy and tomato allergy can induce a potato allergy. Potato specific IgE is thought to worsen Atopic Dermatitis (A.D.). Latex-fruit syndrome is a recognised entity. Potatoes belong to the **plant family** called Solanacae which includes tomatoes and aubergines. They were once thought to be an **evil plant** because of the bad reactions they can cause.

The IgE antibody to patatin protein (sol- t-1) causes severe allergic disease. Patatin is the **primary allergen** in potato. It is heat stable so cooked and raw potato can be an allergenic food in infants. There is a high rate of immune system cross-reactivity with other fruits and vegetables.

Unfortunately the act of frying potatoes (such as chips and crisps) increases the concentration of **Glycoalkaloid (G.A.)**

FOODS THAT CAUSE MIGRAINE

chemicals in potatoes. Other **harmful chemicals** are also thought to be increased by frying – degradation products, triglycerides, mutagenic chemicals , methanol and sometimes Salmonella Typhimurium strains.

The GA chemicals cause a concentration dependent disruption of **GastroIntestinal (GI) barrier integrity** and increased histological colonic injury and small bowel necrosis in hamsters. This can aggravate a diseased bowel causing worse damage. So the potato contains **many allergens** as well as patatin.

Glycoalkaloids in potatoes are toxic chemicals. The toxic threshold is 1.0mg/kg body weight and is often exceeded. The combination of red meat and potatoes is thought to increase the risk of cancer. The **GA chemicals** cause GI and systemic side effects. They cause cell membrane disruption and Acetylcholine inhibition. These are **powerful poisons** and are present in mandrake, henbane and ripe deadly nightshade berries . There is good scientific reporting of **potato poisoning syndrome.** Potatoes are poisonous if turning green or sprouting (chitting). The potato tuber makes toxic quantities of the alkaloid alpha –solanine. The two common GA chemicals are **alpha-solanine and alpha-chaconine.**

Mashed potatoes can cause **nausea and vomiting** after about 4 hours. One study suggested that potatoes may be a risk factor for Schizophrenia.

Steroidal Glycoalkaloids (SGAs) can damage cell membranes and are teratogenic (can cause **birth defects** in

PREVENTING HEADACHES AND MIGRAINES

mammals , mice, frog, chicken and hamster embryos; craniofacial malformations, growth retardation and Neural Tube Defects (NTDs) have been described). Alpha-chaconine is worse than alpha-solanine at causing birth defects. These toxins are **lethal** to hamsters if injected. The death rate of embryos os 50% and the malformation rate is 25%. Solanidine is a **metabolite** derived from alpha-solanine and is teratogenic.

Solanine toxicity is clearly recognised and can be fatal because of effects on the heart. Rat heart calls had increased contraction frequency at 40mcg/ml and stopped beating at 80 mcg/ml, causing death.

The newer genetically modified (GM) types of potato may be less toxic containing lower concentrations of these toxins.

Clearance of these toxins from the body takes over 24 hours so they can accumulate if one eats potatoes every day.
The biological half life of alpha –solanine is 11 hours.
The biological half life of alpha –chaconine is 19 hours.

The **moulds** present on the **dust covering potatoes** can also cause a hypersensitivity pneumonitis (inflamed lung), cough, breathlessness, chest tightness, wheezing, reduced Lung Function Tests (LFTs) such as FVC and FEV1; lung pains and hoarseness. It is reported that the common fungi found on the surface of potatoes are Penicillium species, and Aspergillus fumigatus. There is a high rate respiratory reactions to such organic dust exposure.

It is important to avoid eating blighted diseased potatoes.

FOODS THAT CAUSE MIGRAINE

The potato tuber itself may have **fungi within it** such as Fusarium Solani which are toxic to the liver and kidneys.

One report described that poorly cooked potato salad on a ship's buffet caused **food poisoning** symptoms for 3 days afterwards.

Animal tests on **pesticide** (Croneton) used on potatoes in Russia caused dysfunction of the liver and vascular system in rats. Another pesticide is Ethoprop which is used to kill nematode worms. It is advisable to try potatoes labelled as **"organic" or "natural"** to avoid such chemicals if one suffers from headaches.

The written scientific studies show that other foods also contain **plant toxins:**
- maize contains high phytate levels
- rape contains glycosinolates
- tomatoes contain tomatins, solanine, chaconine, lectins and oxalate
- potatoes contain solanine, chaconine, protease inhibitors and oxalate
- soya beans contain protease inhibitors, lectins, isoflavones, phytate.

One study showed high levels of **lead and cadmium** in potatoes in Poland.

CHEMICALS IN OTHER FOODS
The studies on plant proteins indicate that they can cause headaches. Tyramine and beta-phenyl-alanine sensitivity can cause **headaches**. Hypersensitivity reactions to avocado,

PREVENTING HEADACHES AND MIGRAINES

bananas, kiwi fruit, and potatoes can cause **headaches**. These all come under the umbrella of **food intolerance** reactions.

It is recognised that foods that are naturally high in **histamine content** such as chillies and peppers can cause headaches after they are eaten.

Positive skin prick tests are reported to celery, carrots, birch and mugwort pollens because of immune system cross reactions. Some RAST tests are also positive.

Immune system reactions **(food allergy)** to beef, milk, potato, fish and eggs may cause Pancreatitis. Food allergens are also thought to rarely cause Nephrotic Syndrome by stimulating the immune system.

It is described that the **sulphites** on lettuce, shrimp, dried apricots, white grape juice, mushrooms and mashed potatoes can cause asthma attacks. So it is important that all such foods are **well cooked – a good method is to slow boil them with some mild herbs or spices for 45 minutes.**

BANANA RESEARCH STUDIES
Bananas contain **phenolic acids,** polyphenol oxidase, antioxidants and flavinoids. The stems also contain phenolic acids. Unripe green bananas contain phytosterols. Please see the **references** listed at the end of this book.

If a person has a **diseased bowel** and a damaged GI barrier, then I think these phenol type chemicals will not be broken down, going into the bloodstream and affecting the brain

FOODS THAT CAUSE MIGRAINE

causing intense drowsiness and headache.

COELIAC DISEASE (C.D.)

This is a bowel disease which destroys the lining of the small intestine. The fundamental problem here is an immune system reaction to **gluten** present in **wheat, oats, barley and rye**. So this is a **true food allergy** involving an immune system reaction which damages the lining of the small intestine.

When such an ill person eats gluten in foods, one of the first symptoms are **headache, muscle pains, fatigue** within the first 6 hours, as a side effect of the immune system reaction from the gut. If a large dose of **gluten** is eaten then this can be a severe pulsating throbbing **migraine** (cerebral vasculitis; temporal arteritis; vascular headache).

There are already many good **Gluten Free (G.F.) cookbooks** on the internet. The food industry uses gluten widely to bind foods together. It is a glue like substance.
So **gluten is often added** to sauces, ketchup, baking powder, soy sauce, beers & other types of alcohol, barley water, malted milk drinks, mustard products, stuffing mixes, canned foods and stew, sausages, some burgers etc.. Most of the cookbooks have sections describing this. Some people with headaches are not aware that they are **reacting to gluten** or wheat which has been added to these foods.

The UK supermarkets these days have a good range of **Gluten Free (G.F.) foods** called the **"Free From" range** created by expert dieticians. The **bigger branches** have a wider range of GF frozen pizzas, rice noodle meals, chicken steaks, macaroni

PREVENTING HEADACHES AND MIGRAINES

cheese etc.; there are also goof GF pasta, spaghetti ,chocolates, cereals and snacks. It is worth trying these and keeping a symptom diary of headaches. Not all the products are "organic" or "natural" so some people may still react to the **food additives (chemicals), baking soda or even the chocolate protein** in some products, causing a headache.

The **"Free From"** sections contains **oat** based cereals and museli but this is puzzling because oats can contain gluten. Stomach cramps , nausea and heartburn have been reported after eating these.

Some cereals are made from **Buckwheat** but sensitive ill people with a diseased bowel can react to this too (headaches).

Water, tiny amounts of diet cola drinks and milk are usually alright for these ill people. Fizzy drinks with artificial chemical **food additives** can cause headaches because these are not broken down by the diseased bowel. Tea and coffee can cause drowsiness – likely effect of compounds on the brain.

Most waiters in **restaurants** are well informed and most has **Gluten Free (GF) options in their menu.** This is common in Italian restaurants which have GF Pizza or Pasta. Asian restaurants have **rice based foods** which are GF in type but one needs to avoid very spicy curries because of the histamine content. It is worth phoning the restaurant before going and planning ahead to prevent a bad migraine. If one suffers from headaches it is advisable to try a diet that is **Gluten Free (GF) and not spicy.**

FOODS THAT CAUSE MIGRAINE

Spasm of scalp muscles & arteries after exercise, housework, foods, bright sunlight, eye strain etc. causes bad Tension Type Headaches (TTH) or Migraines. There is an overlap between these two categories (types) of headaches. Fast exercise is more likely to cause artery spasm so exercise at medium speed, not fast speed.

Spasm of scalp muscles and arteries after exercise, housework, foods, bright sunlight, eye strain etc. cause Tension Type Headaches (TTH) or Migraines. There is an overlap between these 2 categories (types) of headaches. The temporal muscles (of the temple area) of the skull are often affected.

A throbbing pounding pulsating headache indicates artery spasm, often after certain foods or fast exercises. Fast exercises can cause a headache/ migraine 1 or 2 days later. This diagram shows the muscles of the scalp area. Excessive pulling or stretching of the scalp muscles from heavy lifting at work, weight training, gymnastics, yoga or karate stretches can cause bad headaches 1 to 2 days later.

PREVENTING "POST EXERCISE MIGRAINE" (PEM)

These guidelines are based on very careful health diaries over the past 8 years. Written research studies indicate that one third of people get post exercise headaches or migraines. Many peoples have **"sensitive arteries" (and sensitive muscles)** in the forehead, scalp or skull making them prone to pulsating throbbing artery-spasm or muscle-spasm migraines. Such people **should not over-exercise** as this causes me to have a Post-Exercise Migraine (PEM) or Post Exercise Headache (PEH), often of delayed onset, starting 1 or 2 days after the exercise session, housework or other physical activity.

We all recognise that after jogging, our leg muscles are swollen, stiff and painful for 1-3 days afterwards. The technical medical term is **"exercise induced myositis"** i.e. inflammation of muscles after exercise. The muscle fibres (sarcomeres) actually become damaged, swollen & painful. This is also true for the **neck, shoulder head, forehead & scalp muscles**. There are many written research studies confirming this (see the "References" listed at the end of this book). Thus after **"eccentric exercise "** (strenuous, unaccustomed exercise) using the neck, shoulder scalp or head muscles , one can suffer a headache for 2-3 days

PREVENTING "POST EXERCISE MIGRAINE"

afterwards e.g after swimming, rugby, soccer, hockey, cycling, using a rowing machine or cross-trainer, doing sit-ups, push-ups, weight training, crunches , heavy lifting at work, carrying a heavy ruck-sack or heavy shoulder bag, using a screw driver or other manual equipment vigorously. All of these activities pull & stretch the forehead, scalp, neck & shoulder muscles **causing muscle damage, pain & swelling for a few days afterwards**. Repeated bouts of damage to the same muscles will worsen the headaches over the years. Genetic predisposition, family history, age & gender are also relevant considerations.

There are several raised chemicals in the blood (cytokines) after such eccentric exercise reflecting an acute inflammatory reaction of the muscles. There is also a **Delayed Onset Muscle Soreness (DOMS)** observed in mice & probably also occurs in humans. The white cell count (WCC) and Adrenaline (Epinephrine) levels are also raised. The immune system cells are also affected. Such **eccentric exercise** is a common trigger factor for headaches/ migraines , and the headaches/ migraines can last for 1-3 days after such exercise.

It is puzzling & paradoxical that a food may cause a migraine on day 1 but not on day 2. This may be explained by the fact that the forehead & scalp muscles are damaged & more swollen on day 1 than on day 2 after eccentric exercise (trigger factor 1). The chemicals in some foods then worsen the underlying inflammation, especially if they contain histamine (trigger factor 2). The underlying blood chemicals are also more abnormal after eccentric exercise on

PREVENTING HEADACHES AND MIGRAINES

day 1 than on day 2. **Two or three trigger factors are often needed to produce a bad migraine of 100% severity.**

Many people admire people in the army who seem to be able to exercise all day long without problem.

Research studies on teenagers & university students show that one third of them suffer from Tension Type Headaches (TTH) or Migraine after weight training, rugby or other strenuous exercise.

Over exertion can cause a headache or migraine up to 2 days afterwards.

AVOID fast exercises as this can cause a migraine 1 or 2 days later (greater effect on the heart & circulation).

It is a good idea to **rotate the muscle groups** when exercising; so one day exercise mainly upper body muscles and the next time use mainly leg muscles. This helps prevent headaches.

PREVENTING "POST EXERCISE MIGRAINE"

Fast exercises (e.g. running) seem to hyperstimulate the circulation causing pounding throbbing headaches in timing with the heartbeat, often 1 or 2 days after exercise, possibly when the circulatory system is trying to recover from the strenuous activity.

The scalp, neck, head & forehead muscles & arteries are pulled & stretched during rowing. This can cause Tension Type Headaches (TTH) or Migraines 1 or 2 days later.

PREVENTING HEADACHES AND MIGRAINES

AVOID FAST EXERCISES (e.g. fast leg changes, star jumps, open-close, fast running, fast running on the spot) as these seem to hyper-stimulate the circulation (arteries) causing an artery-spasm migraine (of 40% severity) 1 or 2 days later. So do exercises at **medium speed** (moderate speed), not a very fast speed.

DO NOT do fast running or fast rowing or other fast exercises, fast housework, or other fast physical activity

DO run at slow or medium speed, because this does not usually cause a headache afterwards. Slow jogging for 5 minutes is a good warm-up for an exercise session.

PREVENTING "POST EXERCISE MIGRAINE"

Patient "I" found that if she jogged 2 kilometres he got a headache of 40% severity 1 or 2 days later, but jogging just 1 kilometre was okay for her. She observed this several times & then decided to just jog 1 km as an "exercise session".

AVOID doing too much weight training. During weight-lifting or heavy work:
 a) the head, scalp & forehead muscles are pulled & stretched
 b) one strains & grunts (**valsalva manoevre**) to lift the heavy object.
 c) The circulatory system is hyperstimutated

These three factors combine to cause a bad Tension Type Headache (TTH) or Migraine headache 1 or 2 days later. This applies if someone lifts very heavy items at work e.g. postbags, furniture, TVs, gardening items, beer barrels, carrying a heavy ruck-sack etc.

One subject found that doing 4 sets of 10 repetitions of curls (weight training) was OK but if he did more that this he got a bad headache 1 or 2 days afterwards. When he increased his weight training in 1993 to 100 repetitions, the headaches worsened in direct correlation to the increased repetitions (reps.) of weight training he performed. This occurred several times so he decided continued doing just 40 repetitions. After 15 years he finally realised that **tomato ketchup/ sauce and salad creams containing gluten worsened his migraines**. So he stopped all such sauces (probable high histamine content) and was able to increase his weight-training. As previously mentioned in the introduction many sportsmen & sportswomen need 9.5 to 11

PREVENTING HEADACHES AND MIGRAINES

hours sleep after such strenuous exercise sessions (instead of the usual 8 hours sleep).

He noticed that his worst migraine ever was 1 or 2 days after **digging up the garden**. The action of lifting the gardening fork and shovel pulled & stretched his forehead & scalp muscles & arteries. It seemed as if the headaches occurred 1 or 2 days later when his muscles & arteries were trying to recover from the excessive pulling & stretching. This headache was far worse than after weight-training because his forehead & scalp was being pulled to a greater extent. The pains were worse on the right hand side of his head because he was right-handed & placed greater strain on this side of the body, pulling the forehead & neck muscles to a greater degree on this right side. He was a postman & noticed that he had worse right sided head pain & migraine for 1-2 days after carrying heavy postal sacks on this right shoulder .

Gentle body & spine stretches can be done every day but it is best to do **strenuous spinal stretches** just 2 to 3 times per week because they can cause the spine & low back muscles to be **swollen, stiff and painful for 2 to 3 days afterwards (exercise induced myositis).**

PREVENTING "POST EXERCISE MIGRAINE"

BEWARE excessive weight lifting or digging the garden.

AVOID doing more than 30 PUSH-UPS as these cause migraines 1 or 2 days later due to the straining **(valsalva manoevre)**. Push-ups on the knuckles (in Karate classes)

PREVENTING HEADACHES AND MIGRAINES

seem to be especially bad for sensitive people causing migraines 1 or 2 days later, worsened by eating migraine inducing foods.

The written research studies indicate that performing a **"Valsalva manoevre"** (holding our breath) when we strain or lift items can cause a headache afterwards, probably from increased pressure inside the skull as well as increased pressure in the skull arteries & muscles.

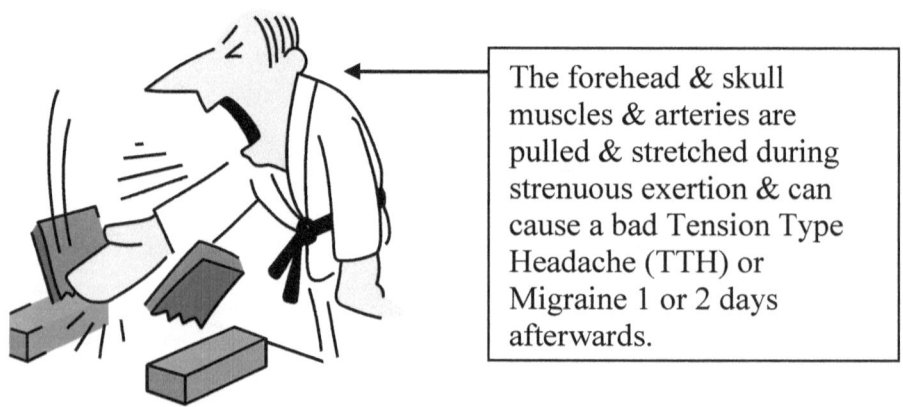

The forehead & skull muscles & arteries are pulled & stretched during strenuous exertion & can cause a bad Tension Type Headache (TTH) or Migraine 1 or 2 days afterwards.

Carrying heavy items on one day may cause a delayed onset vascular type headache 1-2 days later.

PREVENTING "POST EXERCISE MIGRAINE"

AVOID CRUNCHES & SIT-UPS as these cause excessive pulling of the frontal skull arteries & muscles causing a migraine 1 or 2 days later. Doing excessive sit-ups can also cause abdominal wall muscle pains.

Single or double "leg raises" are good for the abdominal muscles & do not cause a headache afterwards but may cause low back pain.

PREVENTING HEADACHES AND MIGRAINES

AVOID FRONT ROLLS, GYMNASTICS OR SOMERSAULTS as these can cause migraines 1 or 2 days later because of :
  a) the gravitational movement of brain fluid & blood
  b) the pulling & stretching of the forehead, scalp & neck muscles & arteries.

AVOID THE HEAD-DOWN POSITION for any prolonged time, because the downward effect of gravity on the brain, blood vessels & scalp muscles can cause a headache. The head-down tilt (position) can cause engorgement & swelling of the blood vessels inside the skull & scalp.

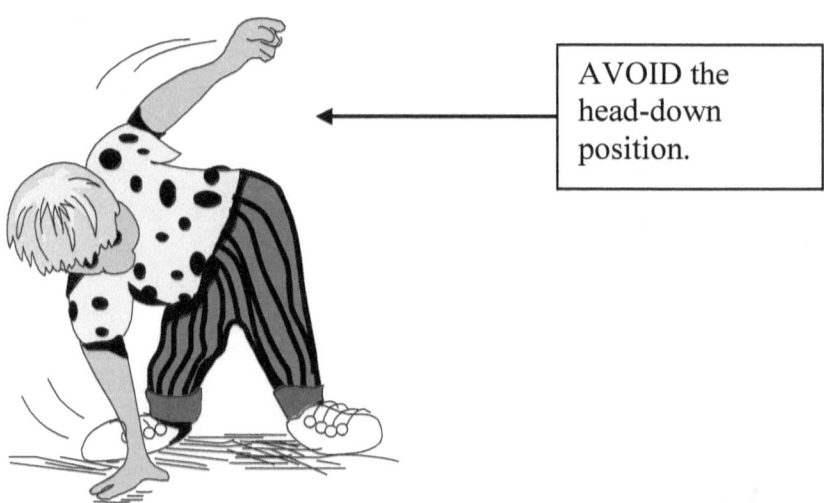

AVOID the head-down position.

Exercises that pump-up the chest, arm & shoulder muscles making them feel swollen & tense can cause bad **Tension Type Headaches (TTH)** or Migraine 1 or 2 days later e.g. avoid breast stroke when swimming, excessive heavy weight lifting etc.

PREVENTING "POST EXERCISE MIGRAINE"

AVOID EXCESSIVE SETS OF YOGA OR KARATE STRETCHES as these pull the forehead & head muscles & arteries excessively, causing a migraine 1 or 2 days later. There is **exercise induced myositis** and it seems like the arteries go into spasm when they try to recover from the stretching, 1-2 days later. Do ONE SET only of these stretches (usually needing about 15 minutes) & then stop.

> The forehead, scalp & neck muscles & arteries are pulled & stretched excessively during Yoga, Gymnastics or Karate stretching

AVOID doing squats or half squats or Kung Fu squats (archer's stance) because the excessive straining (valsalva manoevre) can cause a migraine (of 40% severity) 1 to 2 days later.

One subject found that tiny amounts of fudge or chocolate caused bad migraines up to 12 hours later, spoiling his whole day. So he stopped all chocolates & sweets. This made him feel a bit sad but at least he had far less migraines. He found that **wheat products containing gluten & chocolate products** were especially bad. It was a sobering thought that wheat caused his migraines up to 16 hours after

PREVENTING HEADACHES AND MIGRAINES

eating it because the food stayed in the **gut reservoir** for this duration of time.

When people **keep a health diary** of headaches & migraines, our understanding improves & changes month by month e.g. for 10 years one subject felt that he had just post-exercise headaches 1-2 days after weight training and Karate exercises. He then realised that these were worsened by certain foods & a friend said that he probably had migraine. For a while, he reflected that he had purely food-induced migraine but found that this view was incorrect. He concluded that he got a 40% migraine 1-2 days after certain exercise sessions (e.g. weight training, Karate), worsened to a migraine of 100% intensity **after chocolates, burgers, salad creams, tomato ketchup & gluten other foods.** No two people will have an identical experience. Keeping an accurate health diary, migraine diary, exercise diary & food diary, for months & years, is a big challenge. This is a long term learning process for each person.

He did a sedentary office job & found that if he did no proper exercise sessions for a week, **he got a migraine from lack of exercise.** This was a pulsating throbbing "circulation type" (artery spasm) migraine from under-exercise. It was like having "withdrawal symptoms" from lack of exercise & lack of physical activity. This realisation was an unexpected shock. Once he understood this, **he ensured that he did 20 to 30 minutes of moderate exercise 2 to 3 times per week** e.g. on Mondays, Thursdays & Sundays. This is consistent with the **UK Allied Dunbar National Fitness Survey (AD NFS) exercise protocol** which specifies 12 exercise sessions per month (e.g. 3 times per week ,at medium speed or fast

PREVENTING "POST EXERCISE MIGRAINE"

speed). He read in a book on migraine that regular moderate exercise (without overdoing it) helped prevent migraine."

Sensitive people should AVOID doing exercise of strenuous physical activity 2 days in succession e.g. do NOT do weight-training the day after washing the car, vacuum cleaning, scrubbing the bathtub or moving heavy items. These types of housework already pull & stretch the head muscles & arteries. If one needs an exercise session the day after such activity one should do **simple jogging for 10 minutes** (session 1) below (fitness session). Washing the car counts as an "**exercise session**" as does moving furniture or doing DIY. These count as sessions of "**physical activity**". Do not over exercise. Having an exercise session 2 to 3 times weekly is optimal e.g. on Mondays, Thursdays & Sundays with 2 days rest between "exercise sessions". Some people do not seem to realise that they are doing 10 or 15 "exercise sessions" per week.

In general one should also have a low salt diet and **low histamine diet** avoiding chillies, peppers, spices, raw tomatoes, raw strawberries, canned fish, chemical food additives, avoiding monosodium glutamate (MSG) etc.. These foods can cause migraines 1 –12 hours after eating them. **A low diary, low wheat & low nut diet** is also advised to calm the immune system & reduce migraines. This is especially important in the first 24 hours after exercise. One can get a severe (100%) migraine just from a bad meal alone. Avoiding such foods may mean you get a moderate (50%) migraine or just a mild migraine (30% intensity). This makes life a lot easier to bear.

PREVENTING HEADACHES AND MIGRAINES

If he had a migraine from lack of exercise ,doing a **moderate weight-training session** (as mentioned below) worked like magic and got rid of the migraine. It was as if the weight-training adjusted the muscle tension in the upper body (neck, shoulder & scalp muscles), to relieve the Tension-Type Headache (TTH) or migraine. Other types of exercise session were not effective at getting rid of the pain.

SUGGESTED EXERCISE SESSIONS TO TRY

1. FITNESS SESSION
Slow jogging around the room 50 times (1 kilometre) as a good session and doing 2 sets of 20 leg raises. Do NOT run 2 km because this is excessive & you might get a post-exercise migraine from **hyperstimulation of the circulation** for 1 or 2 days afterwards (of 40% severity). Total duration 10 to 15 minutes.

2. WEIGHT TRAINING
Five minutes of **slow jogging** + 2 sets of 20 leg raises as a warm-up. Then, 40 repetitions (reps.) of weight training (10 reps x 4 sets) using 20 or 25kg on the long barbell (or up to 40% of your body weight). Total duration of about 15 minutes.

If you do 50 reps. you may get a migraine 1 or 2 days later from **exercise induced myositis** (muscle damage to neck, shoulder, head & forehead muscles (over exercising)). If you use 3x 10 reps. using a heavier weight (30kg) on the bar you may also get a migraine 1 or 2 days later (over exercising). Some people may be okay with doing 100 reps. per session. This varies for each person.

PREVENTING HEADACHES AND MIGRAINES

3. KARATE, YOGA OR KUNG FU SESSION
ONE set only of Kung Fu, Karate or yoga stretches + **gentle slow jogging**, followed by kicking OR movement patterns .If you do two sets of yoga stretches or excessive exercise in this session you may get a migraine 1 or 2 days later from **exercise induced myositis**. Total duration of about 40 minutes.

4. EXTRA STRETCHING SESSION
Two sets of Yoga or Karate STRETCHES (including the splits x2). You can do **gentle slow jogging** for 2-3 minutes & 4 sets of leg raises. AVOID fast running with this session as this can cause a migraine the next day (over- exercising). Total duration of about 40 minutes.

MEDICATION FOR HEADACHES & MIGRAINES

Acute Treatment of Migraine:
Most chemists in the UK have a range of good pain killers available "over the counter" (OTC), without needing a prescription for the "Acute treatment" of a migraine e.g. Paracetomol, Cocodamol, Ibuprofen, Aspirin, Syndol, Migraleve, etc. Also , Sumatriptan (Imigran) is available on prescription, from the GP. These medicines can be used in addition to the "preventive" tablets mentioned below.

Prevention of migraine:
If someone suffers regular migraines the GP can prescribe some tablets to stabilize the arteries & so prevent artery (circulatory) spasm, thereby preventing migraines e.g. Propranolol, Pizotifen (Sanomigran). Hospital specialists may also advise other additional "preventive" tablets e.g. Gabapentin, Amitriptyline.

EYE STRAIN

Prolonged concentrating on a computer screen, sewing, knitting, video games or driving (e.g. for over 1 hour) may cause straining of the eye muscles as they try to focus. This is especially so if you have defective eyes (short sight or long sight). This is mentioned **the book "Understanding migraine" listed in the references below**.

Person "H" noticed that when he took off his glasses, closed his eyes & lay down to rest, his migraine reduced by about 40% within 30 minutes. He also found that having a warm bath or shower eased the throbbing migraine by 20% (probably because the warmth had an effect on the arteries of the body (circulatory system)).

REFERENCES:

Here is a list of books & written research studies **referred to** in the compilation of this book. Most libraries can obtain the **written research studies** for you from the publishers of specialist scientific journals when you pay a fee. The books are available in many libraries of can be ordered from your local book shop or websites e.g. Amazon.co.uk .

The computers in a "science library" can **search a topic** for you based on the **"string of letters or characters"** e.g. you can ask for a **quick literature search** on research studies containing the words **"headache"** and **"chocolate"** in the English language, over the past 10 years; or the words **"headache"** and **"exercise"** , **"exercise induced myositis", the effect of salt on the heart & circulation"** etc. Ask the clever librarians about the topics you wish to learn about. Most libraries will charge a fee for doing this. If you have paid to be a member of a scientific organisation e.g. the **British Medical Association (BMA) or American Medical Association (AMA),** you can just phone up the library of your huge organisation & request a literature search on any topic you wish to learn about. This can be a health topic or any other topic you choose e.g. politics, marketing etc. This is a very useful resource.

PREVENTING HEADACHES AND MIGRAINES

The "Medline" database of research studies is widely used by many doctors. The UK "Cochrane" subject searches are also widely used.

For reasons of copyright I am not permitted to reproduce extracts or abstracts of these articles here so I have given the good articles a 2 star rating (**) or a 3 star rating (***) below.

1. **"The Allergy Bible"** by Linda Gamblin (Consultant Professor Jonathon Brostoff). First published in 2001. ISBN 1 902 757 54 8.
   This a very good book which has a good information on Food Intolerances. At the back it has details of low histamine diets, low amine diets etc. ***

2. **"The Complete Guide to Food Allergy and Intolerance"** by Professor Jonathon Brostoff and Linda Gamblin. *** ISBN 0-7475-3430-6.
   Another excellent book.

3. **"E for Additives"** by Maurice Hannsen & Jill Marsden. First printed in 1987. ISBN 0 7225 1562 6. This is a good book containing details of food additives & substances. It lists the biological side effects of many additives. There is good research evidence for avoiding pesticides, colourants, preservatives, artificial flavours & other additives. We live in an advanced society where foods are very processed & have additives added. ***

REFERENCES

4. **"Understanding Migraine"** by Dr Marcia Wilkinson & Dr Anne MacGregor. First printed in 1993. ISBN 1 898205 04 3. This is an excellent small book about migraine published by the British Medical Association (BMA). ***

5. **Foods causing migraine**:
   a) Leira, R.Rodriguez. Diet and migraine (Spanish).Revista de Neurologica, 24 (129): 534-8, 1996 May. This very good article has a long list of foods & additives that cause migraine. ***
   a. Vaughan TR, The role of food in the pathogenesis if migraine headache. ** Clinical Reviews in Allergy, 12 (2) :167-80, 1994. (Tyramine can cause food induced migraines).
   b. Novembre, E.Dini et al., Unusual reactions to food additives (Italian). *** Pediatria Medica e Chirurgica, 14 (1): 39-42, 1992, Jan – Feb. (Tartrazine & benzoates can cause migraines).
   ***
   c. Diamond, S. et al. Diet & headache. Is there a link? Postgraduate Medicine, 79 (4):279-86, 1986 March.
   d. S. Seltzer. Foods, and food and drug combinations responsible for head and neck pain. Cephalalgia 2(2):111-24, 1982 June. ***
   e. Gibb CM et al., Chocolate is a migraine provoking agent. Cephalalgia. 11(2):93-95, 1991 May.

f. Applebaum, R.S., Diet and migraine. Journal of the American Dietetic Association. 84 (8):942, 1984 August.
g. Schweitzer JW et al.. Chocolate, beta-phenethylamine and migraine re-examined. Nature. 257 (5523):256, 1975 Sep. 18.
h. Hanington, E. Monoamine oxidase and migraine. Lancet. 2(7889): 1149-9, 1974 Nov 9.
i. Moffett, A.M. Effect of chocolate in migraine: a double-blind study. Journal of Neurology, Neurosurgery & Psychiatry. 37(4):445-8, 1974 April.
k) Blau JN & Diamond S.Dietray factors in Migraine precipitation: The physician's view. Headache. Vol. 25(4):184-87, 1985.***
l) Evans RW. Crying migraine. Crying can cause migraine. Headache 38(10): 799-800, 1998 Nov-Dec. **
m) Misek B et al. Salad bar preservatives can be headache triggers. NHF Headlines 2000 Mar-Apr; (113); 9. ** (So avoid salad creams, burgers & coleslaw).
n) Furniss S., Wine, headache, migraine and additives. Immunology & Allergy in Practice Vol. 12(4) :128-132, 1990
o) Van den Eaden SK et al. , Aspartame ingestion & headaches. Neurology 44(10):1787-93, 1994 Oct.
p) Scher W, A possible role for nitric oxide in glutamate (MSG)-induced Chinese restaurant syndrome, glutamate induced Asthma, "hot-dog headache", pugilistic Alzheimer's disease and

REFERENCES

other disorders. Medical Hypotheses. 38 (3): 185-8, 1992, July
q) The book "The Great Wine Swindle" by Malcolm Bluck. (2008). Published by Gibson Square.

6. **Headaches and posture:**
a) Giacomini PG et al. Impaired postural control in patients with Tension-Type Headache (TTH). European Journal of pain. Ejp8(6):579-83, 2004 Dec.
b) Chou CH et al. Cephalic venous congestion aggravates only migraine-type headaches. Cephalalgia. 24 (11): 973-9, 2004 Nov. (Increase cerebral venous pressure or intracranial pressure when lying down). Avoid the head-down position. ***
c) Hannerz J. et al. Head-down tilt provocation in patients with chronic Tension –Type Headache (TTH) and controls. ** Headache. 44 (3):223-9, 2004 March. Headache 44(2): 154-9,2004 Feb.
d) Mokri B. Low PA. Orthostatic headaches without CSF leak in postural tachycardia syndrome. Neurology, 61(7):980-2, 2003 Oct 14.
e) Ferrante E. et al. Postural headache in a patient with Marfan's Syndrome. Cephalalgia. 23(7): 552-5, 2003 Sep.
f) De Marinis M. Migraine and autonomic nervous system function : a population based, case-control study. Neurology. 61 (3) :424-5, 2003 Aug12.
f) Yousry I et al. Cervical MR imaging in postural headache. American Journal of Neuroradiology. 22(7): 1239-50, 2001 Aug.

PREVENTING HEADACHES AND MIGRAINES

g) Gonzalez DP. Lumbar Puncture Headache exacerbated by recumbent position. Military Medicine. 165 (9): vi, 690,2000 Sep.
h) Chen WT et al. Cerebellar Haemorrhage presenting as orthostatic Headache. Neurology 53(8): 1887-8, 1999 Nov10.
i) Schievink WI et al. Nonpositional headache cause by spontaneous intracranial hypotension. Neurology 51(6):1768-9, 1889 Dec.
j) Arjona A.et al. Orthostatic headache. Headache 38(2): 142-3, 1998 Feb.
k) Mokri B et al. Syndrome of Orthostatic headaches. ***Mayo clinic proceedings. 72(5): 400-13, 1997 May.
l) Kruszewski P et al. Cluster Headache : cardiovascular response to head-up tilt. Headache. 35 (8): 465-9, 1995 Sep.
l) Martin R. et al. Duration of decubitus position after epidural blood patch. Canadian Journal of Anaesthesia. 41 (1): 23-5, 1994 Jan.

7. **Medication causing headaches:**
a) Palacio E. et al. Topiramate responsive headache due to idiopathic intracranial hypertension in Behcet syndrome. Headache 44 (5):436-7, 2004 May
b) Rozen TD, Worsening of headaches on Topiramate. A low CSF pressure syndrome? Headache 43(7): 819-20, 2003 Jul-Aug.
c) Frediani F., Anticonvulsant drugs in primary headache prophylaxis. Neurological Sciences 25 Suppl3 :S161-6, 2004 Oct.

REFERENCES

d) Hershey AD et al., Effectiveness of Topiramate in the prevention of childhood headaches. Headache 42 (8): 810-8, 2002 Sep.

8. **Post exercise headache:**
   a) Koseoglu et al. Aerobic exercise and plasma beta endorphin levels in patients with migrainous headache without aura. ***Cephalalgia. 23 (10(: 972-6, 2003 Dec.
   b) Turner J., Exercise –related headache. (Exertional headache). Current Sports Medicine reports. 2 (1):15-7, 2003 Feb. ***
   c) Imperato J., Benign exertional headache .** Annals of Emergency Medicine. 41 (1): 98-103, 2003 Jan.
   d) Sjaastad O. et al. Exertional headache. Vaga study of headache epidemiology. Cephalalgia. 22 (10): 784-90, 2002 Dec.
   e) Kinart CM et al. Prevalence of migraines in NCAA division 1 male and female basketball players. National Collegiate Athletic Association. Headache. 42(7): 620-9, 2002 Jul-Aug.
   f) Jull G. et al., RCT of exercise and manipulative therapy for cervicogenic headache. Spine. 27(17): 1835-43; discussion 1843, 2002 Sep.
   g) Wiehe M. et al., Migraine is more frequent in individuals with optimal and normal blood pressure; a population based study. Journal of Hypertension. 20 (7):1303-6, 2002 July.
   h) Ashina M. et al., In vivo evidence of altered skeletal muscle blood flow in chronic tension –type headche. Brain. 125(Pt2): 320-6, 2002 Feb.

PREVENTING HEADACHES AND MIGRAINES

i) Burtscher M. et al., Effects of Aspirin during exercise on the incidence of high-altitude headache. Headache .41 (6):542-5, 2001 Jun.

j) Razavi M. et al., Hemiplegic migraine induced by exertion. Archives of Neurology. 57 (9): 1363-5, 2000 Sep

k) Lane JC, Migraine in the athlete. Seminars in Neurology .20(2): 195-200, 2000

l) Vingen JV et al., Sensitivity to various stimuli in primary headaches. Headache. 29(8):552-8, 1999 Sep.

m) Williams SJ et al., Sport and exercise headache: Part 2. British Journal of Sports Medicine. 28(2):96-100, 1994 Jun. 28(2): 90-5, 1994 Jun.

9. Immunological cross-reactions to chillies, peppers, spices:

a) Tucke J. et al., Latex type 1 sensitisation and allergy in children with atopic dermatitis. Evaluation of cross-reactivity to some foods. Paediatric Allergy & Immunology. 10(3):160-7, 1999 (Aug). **

b) Maillard H. et al., Allergy associated with pepper and latex: new cross reaction. Allergie et Immunologie. 27 (8):292-4, 1995, Oct. **

c) Ninimaki A. et al., Skin prick tests and in vitro immunoassays with native spices and spice extracts. Annals of Allergy, Asthma & Immunology. 75 (3):280-6, 1995 Sep. ***

d) Ninimaki A. et al., Spice Allergy: results of skin prick tests and RAST with spice extracts. Allergy. 44(1):60-5, 1989 Jan. ***

REFERENCES

e) Van Toorenenbergen AW. Et al., Demonstration of spice-specific IgE in patients with suspected food allergies. Journal of Allergy & Clinical Immunology. 79(1): 108-13, 1987 Jan. ***
f) Wuthrich B. et al.,Food allergy: the celery-mugwort-spice syndrome. Association with mango allergy? Deutsche Medizinische Wochenschrift. 109(25):981-6, 1984 Jun 22. **
g) Wuthrich B. et al., Acne vulgaris: result of food allergen tests and a controlled elimination diet. Dermatologica . 157(5):294-5, 1978. **
h) Eriksson NE., Food sensitivity reported by patients with Asthma and hay fever. A relationship between food sensitivity and birch pollen allergy and between food sensitivity and acetylsalicylic acid intolerance. Allergy. 33(4); 189-96, 1978 Aug. **
i) Barth GA. Et al., Food intake of patients with atopic dermatitis. European Journal of Dermatology. Vol 11 (3)pp 199-202, 2001.***
j) Ebner C. et al. Characterization of allergens in plant derived species: Apiaceae spices, pepper (Piperaceae) and paprika (bell peppers, Solanaceae). Allergy, supplement. Vol. 53(46), pp 52-54, 1998. ***
k) Jensen- Jarolim E. et al., Bell peppers (capsicum annuum) express allergens (profiling, pathogenesis-related protein P23 and betv 1) depending on horticultural strain. International Archives of Allergy & Immunology. Vol.116 (2) pp103-109, 1998. ***
l) Jensen-Jarolim E. et al., Hot spices influence permeability of human intestinal epithelial

monolayers. Journal of Nutrition. Vol. 128(3) pp 577-581, 1998 ***

m) Rance F. et al., Labial food challenge in children with food allergy. Paediatric Allergy & Immunology. Vol. 8(1) pp 41-44, 1997 **

n) Jautova J. et al., Combined fixed exanthema of medicamentous and alimentary origin. Ceskoslovenska Dermatologie. Vol. 72(3) pp 99-102, 1997 ***

n) Stager J. et al., Spice Allergy in celery-sensitive patients. Allergy: European Journal of Allergy & Clinical Immunology, Vol. 46(6) pp475-478, 1991.

   r) Rajaratnam SS. Et al., Always chew your chillies: a report of small bowel obstruction with perforation. International Journal of Clinical practice 55(2):146.2006 March

10. Exercise Induced Myositis :
a) Peake et.al. "Characterisation of inflammatory reponses to eccentric exercise in humans".(Japan) Exercise Immunology Review 11 :64-85, 2005
b) Shephard, Roy J. "Cytokine responses to Physical Activity, with particular reference to IL6: sources, actions & clinical implications" (Toronto, Canada). Critical Reviews in Immunology. 22(3) :165-82, 2002
c) Malm C. "Exercise induced muscle damage & Inflammation : fact or fiction ?" (Stockholm, Sweden). Acta Physiologica Scandanavia 171 (3): 233-9, 2001 March.
d) Stupka N. et al. "Gender Differences in muscle inflammation after eccentric exercise" (Ontario, Canada). Journal of Applied Physiology 89 (6):2325-32, 2000, Dec.
e) Montain, S.J. et. al. "Impact of Muscle Injury and accompanying inflammatory response on thermoregulation during exercise in the heat" (Massachusetts, USA).Journal of Applied Physiology 89 (3): 1123-30 ,2000 Sep.
f) Pederson , B.K. & Toft, A.D. "Effects of exercise on lymphocytes and cytokines".(Copenhagen, Denmark) British Journal of Sports Medicine, 34 (4) : 246-51, 2000 Aug.
g) Pederson, B.K.at.al. "The cytokine response to strenuous exercise". (Copenhagen, Denmark). Canadian Journal of Physiology and Pharmacology 76 (5): 505-11, 1998 May
h) Tisi, P.V. & Shearman C.P. "The evidence for exercise-induced inflammation in intermittent claudication" (Southampton, UK). European Journal of Vascular & Endovascular

Surgery. 15 (1):7-17, 1998 Jan.
i) Bruunsgaard H. et. al. "Exercise induced increase in serum Interleukin 6 (IL6) in humans is related to muscle damage" (Copenhagen, Denmark). Journal of Physiology 499 (Pt 3): 833-41 Mar 15, 1997
j) MacIntrye D.L. et. al. "Delayed Muscle Soreness. The inflammatory response to muscle injury and its clinical implications". (Vancouver, Canada). Sports Medicine. 20 (1): 24-40, 1995 July
k) Pyne D.B. "Exercise-induced muscle damage and Inflammation : a review" (Canberra, Australia). Australian Journal of Science & Medicine in Sport. 26 (3-4): 49-58, 1994, Sep. –Dec.

11. Decomposition of foods

a) B.le las Rivas et al. "Gene Organisation of the ornithine decarboxylase- encoding region in Morganella morhanii". Journal of Applied Microbiology. 102 (6):1551-60, 2007 June.
b) Winter A.R. et al. "Photodegradation of Natural Organic Matter (NOM) from diverse freshwater sources". Aquatic Toxicology. 84 (2):215-22, 2007 August 30.
c) Lavilla I. et al. "Improved microwave assisted wet digestion (MAWD) procedures for accurate Selenium determination in fish & shellfish by flow injection-hydride generation-atomic absorption spectrometry". Analytica Chimica Acta. 591 (2):225-30, 2007 May 22

REFERENCES

d) Lapidot T. et al. "Lipid Hydroperoxidase activity of myoglobin and phenolic antioxidants in simulated gastric fluid". Journal of Agriculture & Food Chemistry, 53 (9): 3391-6, 2005 May 4.

e) Ludemann V. et al. "Determination of growth characteristics and lipolytic and proteolytic activities of Penicillium strains isolated from Argentinian salami". International Journal of Food Microbiology, 96 (1):13-18, 2004 Oct 1. **

f) Noonpakdee W. et al. "Expression of Catalase gene katA in starter culture Lactobacillus plantarum TISTR850 tolerates oxidative stress and reduces lipid oxidation in fermented meat product". International Journal of Food Microbiology, 95 (2) :127-35, 2004 Sep 1. **

g) Piche L.A. et al "Identification of N-epsilon lysine as the main form of malondialdehyde (MDA) in food digesta". Carcinogenesis, 9(3) :473-7, 1988 Mar. **

h) Dworschak E. "Nonenzyme browning and its effect on protein nutrition". Critical reviews in Food Science & Nutrition. 13 (1):1-40, 1980. ***

i) Sokorski Z. et al. "Protein Changes in Frozen Fish". CRC Critical Reviews in Food Science & Nutrition. 8 (1): 97-129, 1976 Sep. ***

j) Schroder I. " On the chemical evaluation of the protein quality of food after proteolytic decomposition" (In German).Zeitschrift fur Ernahrungswissenschaft. 9(2):180-92, 1969 Mar.(This article appeared on the Medline search but I am unable to read German).

12. Histamine poisoning

a) Miki M. et al. "An outbreak of histamine poisoning after ingestion of the ground saury paste in eight patients taking isoniazid in tuberculous ward". Internal Medicine. 44 (11): 1133-6, 2005 Nov **

b) Vlieg-Boerstra B.J. et al. "Mastocytosis and adverse reactions to biogenic amines and histamine releasing foods : what is the evidence ?" Netherlands Journal of Medicine. 63(7):244-9, 2005 Jul-Aug. ***

c) Kuefner M.A. et.al. "Both catabolic pathways of histamine via histamine-N-methyltransferase and diamine oxidase are diminished in the colonic mucosa of patients with food allergy". Inflammation Research. 53 Suppl 1:S31-2, 2004 Mar.

d) Bijlsma P.B. et al. "Food allergy diagnosis by detection of antigen-induced electrophysiological changes and histamine release in human intestinal biopsies during mucosa-oxygenation". Inflammation Research. 53 Suppl 1:S29-30, 2004 Mar

e) Morimoto T. et al. "Brain histamine & feeding behaviour" Behavioural Brain Research. 124 (2):145-50, 2001 Oct 15.

f) Hall M. "Something fishy :six patients with unusual cause of food poisoning" . Emergency medicine (Fremantle W.A.). 15 (3):293-5, 2003 Jun ***

g) Schwab D. et al. "Enhanced histamine metabolism: a comparative analysis of collagenous colitisand food allergy with respect to the role of diet and NSAID use". Inflammation Research. 52 (4):142-7, 2003 Apr ***

REFERENCES

h) Di Lorenzo G. et al. "Urinary metabolites of histamine and leukotrienes before and after placebo-controlled challenge with ASA and food additives in chronic urticaria patients" .Allergy. 57(12):1180-6, 2002 Dec. ***

i) Lee S. et al. "Clinical application of histamine prick test for food challenge in atopic dermatitis". Journal of Korean Medical Science. 1693):276-82, 2001 Jun **

j) Bodmer S. et al. "Biogenic amines in foods: histamine and food processing". Inflammation Research. 48(6):296-300, 1999 Jun ***

k) Amon U.et al. "Enteral histaminosis": Clinical implications". Inflammation Research. 48(6):291-5, 1999 Jun ***

l) Boutin J.P. et al. "Histamine food poisoning". Sante Publique (Vandoevre-Les-Nancey). 10(1):29-37, 1998 Mar ***

m) Diel E. et al. "Histamine containing food: establishment of a German Food Intolerance Databank (NFID)". Inflammation Research. 46 Suppl 1:S87-8, 1997 Mar

n) Jarisch R. et al. "Wine and headache". International Archives of Allergy & Immunology. 110(1):7-12, 1996 May ***

o) Pace V." Food intolerance and allergy". Panminerva Medica. 37 (2):84-91, 1995 Jun. **

p) Smart D.R. " Scromboid poisoning : A report of seven cases involving the Western Australian salmon,Arripis truttaceus" (histamine poisoning from fish). Medical Journal of Australia 157 (11-12):748-51, 1992 Dec 7-21. ***

r) Molinari G. et al. "Hygiene and health importance of histamine as an unhealthy factor in several food products". Annali d Igiene.1 (3-4):637-46, 1989 May-Aug. ***

s) Sattler J.et al. "Food induced histaminosis as an epidemiological problem: plasma histamine elevation and haemodymanic alterations after oral histamine administration and blockade of diamine oxidase "DAO)". Agents and Actions. 23 (3-4):361-5, 1988 Apr. ***

t) Malone M.H. et al. " Histamine in foods : its possible role in non allergic reactions to ingestants". New England & Regional Allergy Proceedings. 7(3):241-5, 1986 May-Jun. ***

u) Taylor S.L." Histamine food poisoning: toxicology and clinical aspects". Critical reviews in Toxicology. 17(2):91-128, 1986. ***

REFERENCES

Effects of salt (sodium chloride) on the heart & circulation :

a) Safar & Lacolley "Disturbance of macro- and microcirculation : relations with pulse pressure and cardiac organ damage". American Journal of Physiology – Heart & Circulatory Physiology. 293 (1):H1-7, 2007 July **

b) Lima et al. "Effect of lifelong high or low salt intake on blood pressure, left ventricular mass and plasma insulin in Wistar rats". American Journal of the Medical Sciences. 331(6):309-14, 2006 Jun. ***

c) Gronholm et al. "Vasopeptidase inhibition has beneficial cardiac effects in spontaneously diabetic Goto-Kakizaki rats". Eurpoean Journal of Pharmacology. 51(3): 267-76, 2005 Sep 20 ***

d) Weisberg et al. "Pharmalogical inhibition and genetic deficiency of plasminogen activator inhibitor-1 attenuates angiotensin II/ salt induced aortic remodelling. Arteriosclerosis, Thrombosis and Vascular Biology. 25(2):365-71, 2005 Feb. ***

e) Ahn et al. "Cardiac structural and functional responses to salt loading in SHR". American Journal of Physiology – Heart & Circulatory Physiology. 287(2): H767-72. ***

f) Tochikubo and Nishijima "Sodium intake and cardiac sympatho-vagal balance in young men with high blood pressure". Hypertension Research – Clinical and Experimental. 27 (6): 393-8, 2004 Jun. **

g) Endemann et al. "Eplerenone prevents salt induced vascular remodelling and cardiac fibrosis in stroke-prone spontaneously hypertensive rats (SHR) ". *** Hypertension 43(6) :1252-7, 2004 Jun.
h) Ye et al. "Myocardial vasoactive intestinal peptide and fibrosis induced by nitric oxide (NO) synthase inhibition in the rat". Acta Physiologica Scandanavia. 179(4): 353-60, 2003 Dec. ***
j) G.Simon "Experimental evidence of blood pressure-independent vascular effects of high sodium diet". American Journal of Hypertension . *** 16 (12): 1074-8,2003 Dec.
k) Finckenberg et al. "Cyclosporine induces myocardial connective tissue growth factor in spontaneously hypertensive rats on high-sodium diet".** Transplantation.71(7):951-8, 2001Apr15
l) Takatsu et al. "Washout of 1-123 meta-iodobenzylguanidine for assessing cardiac sympathetic activity with progression of hypertension in Dahl salt-sensitive rats". Journal of Nuclear Cardiology.6(2) :204-10, 1999 Mar-Apr **
m) Dibona et al. "Effect of endogenous angiotensin II On renal nerve activity and its cardiac baroreflex regulation". Journal of the American Society of Nephrology.9(11): 1983-9, 1998 Nov. ***
n) Young & Funder "Mineralocorticoids, salt, hypertension: effects on the heart". Steroids.61(4): 233-5, 1996 Apr **

REFERENCES

o) Kishimoto et al. "The heart communicates with the kidney exclusively through the guanylyl cyclase-A receptor : acute handling of sodium and water in response to volume expansion".
Proceedings of the Naional Academy of Sciences of the USA. 93(12) : 6215-9, 1996 Jun11

p) Feron et al. "Influence of salt loading on the cardiac and renal preproendothelin-1 mRNA expression in stroke-prone spontaneously hypertensive rats". Biochemical & Biophysical Research ** Communications. 209(1): 161-6, 1995 Apr 6

q) Sun et al. " ACE & myocardial fibrosis in the rat receiving angiotensin II or aldosterone". Journal of Laboratory & Clinical Medicine.
122(4): 395-403, 1993 Oct. **

r) Frohlich et al. "Relationship between dietary sodium intake ,hemodynamics, and cardiac mass in SHR and WKY rats. American Journal of Physiology.
264 (1 Pt 2) :R30-4, 1993 Jan.***

s) Li et al. "Angiotensin II facilitates tricyclic antidepressant-induced changes in QRS duration in the rat". Journal of Toxicology – Clinical Toxicology.
30(1): 83-98, 1992 **

t) Richardt et al. "Effect of ACE inhibitors on cardiac noradrenaline release". European Heart Journal .
12 Suppl F :121-3, 1991 Dec. ***

u) Yuan & Leenen "Dietary sodium intake and LVH in normotensive rats". American Journal of Physiology.
261(5 Pt 2): H1397-401, 1991 Nov. **

v) Boegehold et al. "Peripheral Vascular Resistance (PVR) and regional blood flows in hypertensive Dahl rats. American Journal of Physiology.
261(4 pt 2): R934-8, 1991 Oct. **

w) Blake et al. "Relation of Obesity, high sodium intake And eccentric LVH to LV exercise dysfunction in essential hypertension". American Journal of Medicine .88(5): 447-85, 1990 May. **

x) Du Calair et al. "Influence of sodium intake on LV structure in untreated essential hypertensives". Journal of Hypertension – supplement .
7(6): S 258-9, 1989 Dec. *

y) Iwase et al. "The effects of sodium loading on cardiopulmonary baroreflexes". Clinical & Experimental Pharmacology & Physiology – Supplement .15: 109-11, 1989. **

z1) Stamler et al. "Cardiac status after 4 years in a trial on nutritional therapy for high blood pressure". Archives of Internal Medicine.
149(3):661-5, 1989 Mar

z2) Saito et al. "LV filling characteristics in sodium-depleted and sodium-loaded patients with mild essential hypertension".Journal of Hypertension – Supplememt .6(4):S134-7, 1988 Dec. **

REFERENCES

Effects of sodium bicarbonate on the heart & circulation:

a) Fang et al. "Effects of resuscitation with crystalloid fluids on cardiac function in patients with severe sepsis". BMC Infectious Diseases .8:50 ,2008-09-06

b) Devin et al. "Managing cardiovascular collapse in severe flecainide overdose without recourse to extracorporeal therapy". Emergency Medicine Australasia .19(2) : 155-9, 2007 Apr ***

c) Recio-Mayoral et al. "The reno-protective effect of hydration with sodium bicarbonate plus N-acetylcysteine in patients undergoing emergency percutaneous coronary intervention :The RENO study". Journal of the American College of Cardiology. 49(12): 1283-8, 2007 Mar27. *

d) Briguori et al. " Renal Insufficiency Following Contrast Media Administration Trial (REMEDIAL) : a randomised comparison of 3 preventive strategies". Comment in Circulation :
2007, Aug 21; 116(8): e310, e311

e) Tsai et al. "Effects on alcohol on intracellular pH regulators and electromechanical parameters in human myocardium". Alcoholism :Clinical & Experimental Research .29(10): 1787-95, 2005 Oct. **

f) Cavusoglu et al. "The prevention of contrast-induced nephropathy in patients undergoing percutaneous coronary intervention".Minerva Cardioangiologica 52(5):419-32, 2004 Oct.

g) Amagase & Okabe "On the mechanisms underlying histamine induction of gastric mucosal lesions in rats with partial gastric vascular occlusion. Journal of Pharmacological Sciences .92(2): 124-36, 2003 Jun. **

h) McKinney at al. "Reversal of severe tricyclic antidepressant-induced cardiotoxicity with intravenous hypertonic saline solution".**
Annals of Emergency Medicine. 42(1):20-4, 2003 Jul

i) Wang & Raymond "The effects of sodium bicarbonate" on thioridazine-induced cardiac dysfunction in the isolated perfused rat heart. **
Veterinary & Human Toxicology.
43(2):73-7, 2001 Apr

j) Wang "pH dependent cocaine-induced cardiotoxicity". American Journal of Emergency Medicine. ***
17(4) :364-9, 1999 Jul.

j) Liebelt "Targeted Management Strategies for cardiovascular toxicity from tricyclic antidepressant overdose: the pivotal role for alkalinization and sodium loading . Pediatric Emergency Care.
14(4):293-8, 1998 Aug.

k) Aiello at al. "Evidence for electrogenic sodium bicarbonate symport in rat cardiac myocytes". **
Journal of Physiology.
512(Pt 1):137-48, 1998 Oct 1

REFERENCES

Effects of Potassium Chloride on the heart & circulation :

a) Chow "Does potassium-enriched salt or sodium reduction reduce cardiovascular mortality and medical expenses?". American Journal of Clinical Nutrition. 84(6). 1552-3 **

b) Chang et al. "Effect of potassium-enriched salt on cardiovascular mortality and medical expenses of elderly men". American Journal of Clinical Nutrition . 83(6): 1289-96, 2006 Jun.**

c) Norris et al. "Potassium supplementation, diet vs pills: A randomised trial in postoperative cardiac surgery patients". Chest.125(2):404-9 ,2004 Feb. **

d) Etheridge et al. "A new oral therapy for long QT syndrome : long term oral potassium improves repolarisation in patients with HERG mutations". Journal of the American College of Cardiology. 42(10): 1777-82, 2003 Nov 19.***

e) Obata et al. "Norepinephrine evoked by potassium depolarisation increases interstitial adenosine concentration via activation of ecto-5'-nucleotidase in rat hearts". Journal of Pharmacology & Experimental Therapeutics .305(2): 719-24, 2003 May. ***

f) Okatani et al. "Melatonin counteracts potentiation by homocysteine of KCL- induced vasoconstriction in human umbilical artery :relation to calcium influx.*** Biochemical & Biophysical Research Communications. 280(3): 940-4, 2001 Jan 26.

g) Mervaala et al. "Improvement of cardiovascular effects of metoprolol by replacement of common salt with potassium- and magnesium- enriched salt alternative".British Journal of Pharmacology. 112(2):640-8, 1994 Jun. **
h) Back et al. " Cortical negative DC deflections following middle cerebral artery (MCA) occlusion and KCL- induced spreading depression: effect on blood flow , tissue oxygenation and electroencephalogram (EEG)". ***
Journal of Cerebral Blood Flow & Metabolism . 14(1):12-9, 1994 Jan.
i) Sei & Glembotski "Calcium dependence of phenylephrine -, endothelin-, and potassium chloride- stimulated atrial natriuretic factor (ANF) secretion from long term primary neonatal rat atrial cardiocytes". Journal of Biological Chemistry. 265(13): 7166-72, 1990 May 5 ***

Studies on Potatoes

1.Unique Identifier 18614268 ,Status MEDLINE
Authors :Dinkins CL. Peterson RK.
Institution:
 Department of Land Resources and Environmental Sciences, Montana State University, Bozeman, Montana 59717-3120, USA.
Title:
 A human dietary risk assessment associated with glycoalkaloid responses of potato to Colorado potato beetle defoliation.
Source

REFERENCES

Food & Chemical Toxicology. 46(8):2837-40, 2008 Aug.

2.Unique Identifier 18614739 ,Status MEDLINE
Authors:
Flood A. Rastogi T. Wirfalt E. Mitrou PN. Reedy J. Subar AF. Kipnis V. Mouw T. Hollenbeck AR. Leitzmann M. Schatzkin A.
Institution:
Division of Epidemiology and Community Health and The Masonic Cancer Center, University of Minnesota, Minneapolis, MN 55454, USA. flood@epi.umn.edu
Title:
Dietary patterns as identified by factor analysis and **colorectal cancer** among middle-aged Americans.[see comment].
Comments : Comment in: Am J Clin Nutr. 2008 Jul;88(1):14-5; PMID: 18614718
Source:
American Journal of Clinical Nutrition. 88(1):176-84, 2008 Jul.

3.Unique Identifier 18161826 ,Status MEDLINE
Authors:
Merget R. Sander I. Rozynek P. Raulf-Heimsoth M. Bruening T.
Institution:
BGFA-Research Institute of Occupational Medicine, German Social Accident Insurance, Ruhr-University, Bochum, Germany. merget@bgfa.de
Title:
Occupational hypersensitivity pneumonitis due to **molds in an onion and potato sorter.**

PREVENTING HEADACHES AND MIGRAINES

Source :
American Journ Journal of Industrial Medicine. 51(2):117-9, 2008 Feb.

4. Unique Identifier 17153152 ,Status MEDLINE
Authors :Lee MR.
Institution: University of Edinburgh, Edinburgh, Scotland.
Title:The Solanaceae: foods and poisons.
Source :
Journal of the Royal College of Physicians of Edinburgh. 36(2):162-9, 2006 Jun.

5. Unique Identifier 15649828, Status MEDLINE
Authors :
Mensinga TT. Sips AJ. Rompelberg CJ. van Twillert K. Meulenbelt J. van den Top HJ. van Egmond HP.
Institution:
National Poisons Control Centre, National Institute for Public Health and the Environment, The Netherlands.
tjeert.mensinga@rivm.nl
Title:
Potato glycoalkaloids and adverse effects in humans: an ascending dose study.
Source:
Regulatory Toxicology & Pharmacology. 41(1):66-72, 2005 Feb.

6. Unique Identifier 10499833, Status MEDLINE
Authors : Christie AC.
Institution:
Emeritus Consultant Pathologist, The Wollongong Hospital, New South Wales, Australia.

REFERENCES

Title :Schizophrenia: is the potato the environmental culprit?.
Source : Medical Hypotheses. 53(1):80-6, 1999 Jul.

7. Unique Identifier 11737675 ,Status MEDLINE
Authors :
Majamaa H. Seppala U. Palosuo T. Turjanmaa K. Kalkkinen N. Reunala T.
Institution :
Department of Dermatology, Tampere University Hospital and University of Tampere, Finland. heli.majamaa@tays.fi
Title :
Positive skin and oral challenge responses to potato and occurrence of immunoglobulin E antibodies to patatin (Sol t 1) in infants with atopic dermatitis.
Source :Pediatric Allergy & Immunology. 12(5):283-8, 2001 Oct.

8. Unique Identifier 12498594 ,Status MEDLINE
Authors:
Milanowski J. Gora A. Skorska C. Mackiewicz B. Krysinska-Traczyk E. Cholewa G. Sitkowska J. Dutkiewicz J.
Institution :
Clinic of Lung Diseases, Medical Academy, Lublin, Poland.
Title:
The effects of exposure to organic dust on the respiratory system of potato processing workers.
Source:
Annals of Agricultural & Environmental Medicine. 9(2):243-7, 2002.

9. Unique Identifier 12487228, Status MEDLINE
Authors : Schmidt MH. Raulf-Heimsoth M. Posch A.
Institution :
Research Institute for Occupational Medicine of the Berufsgenossenschaften, Ruhr University, Department of Allergology/Immunology, Bochum, Germany.
nsmis@neuro.hfh.edu
Title :
Evaluation of patatin as a major cross-reactive allergen in latex-induced potato allergy.
Source :
Annals of Allergy, Asthma, & Immunology. 89(6):613-8, 2002 Dec.

10. Unique Identifier 12479649, Status MEDLINE
Authors :
Patel B. Schutte R. Sporns P. Doyle J. Jewel L. Fedorak RN.
Institution :
Division of Gastroenterology, Department of Medicine, University of Alberta, Edmonton, Alberta, Canada.
Title :
Potato glycoalkaloids adversely affect intestinal permeability and aggravate inflammatory bowel disease.
Source : Inflammatory Bowel Diseases. 8(5):340-6, 2002 Sep.

11. Unique Identifier 12209106, Status MEDLINE
Authors : De Swert LF. Cadot P. Ceuppens JL.
Institution :
Pediatric Allergy, the Department of Pediatrics, the Laboratory of Experimental Immunology, and the Division

REFERENCES

of Allergy and Clinical Immunology, University Hospital Gasthuisberg, Leuven, Belguim.
Title :
Allergy to cooked white potatoes in infants and young children: A cause of severe, chronic allergic disease.
Source :
Journal of Allergy & Clinical Immunology. 110(3):524-35, 2002 Sep.

12. Unique Identifier 11737675 ,Status MEDLINE
Authors :
Majamaa H. Seppala U. Palosuo T. Turjanmaa K. Kalkkinen N. Reunala T.
Institution :
Department of Dermatology, Tampere University Hospital and University of Tampere, Finland. heli.majamaa@tays.fi
Title :
Positive skin and oral challenge responses to potato and occurrence of immunoglobulin E antibodies to patatin (Sol t 1) in infants with atopic dermatitis.
Source : Pediatric Allergy & Immunology. 12(5):283-8, 2001 Oct.

13. Unique Identifier 11736750 ,Status MEDLINE
Authors :
Reche M. Pascual CY. Vicente J. Caballero T. Martin-Munoz F. Sanchez S. Martin-Esteban M.
Institution :
Laboratorio de Inmunoalergia, Hospital Infantil La Paz, Castellana 261, 28046, Madrid, Spain.
Title :
Tomato allergy in children and young adults: cross-

reactivity with latex and potato.
Source : Allergy. 56(12):1197-201, 2001 Dec.

14. Unique Identifier 11642572 ,Status MEDLINE
Authors :
Gomez Torrijos E. Galindo PA. Borja J. Feo F. Garcia Rodriguez R. Mur P.
Institution :Allergy Section, Hospital Complex of Ciudad Real, Spain.
Title : **Allergic contact urticaria from raw potato.**
Source :
Journal of Investigational Allergology & Clinical Immunology. 11(2):129, 2001.

15. Unique Identifier 10828499 ,Status MEDLINE
Authors : Novak WK. Haslberger AG.
Institution :
Institute for Nutritional Sciences, University of Vienna, Austria.
Title :
Substantial equivalence of antinutrients and inherent plant toxins in genetically modified novel foods.
Source : Food & Chemical Toxicology. 38(6):473-83, 2000 Jun.

16. Unique Identifier 10792367 , Status MEDLINE
Authors : Strong FC 3rd.
Institution :
Departamento de Ciencia de Alimentos, Faculdade de Engenharia de Alimentos, Universidade Estadual de Campinas, SP, Brasil.
Title :

REFERENCES

Why do some **dietary migraine patients** claim they get headaches from placebos?.
Source : Clinical & Experimental Allergy. 30(5):739-43, 2000 May.

17. Unique Identifier 10753018 ,Status MEDLINE
Authors :
Seppala U. Palosuo T. Seppala U. Kalkkinen N. Ylitalo L. Reunala T. Turjanmaa K. Reunala T.
Institution : National Public Health Institute, Helsinki, Finland.
Title :
IgE reactivity to patatin-like latex allergen, Hev b 7, and to patatin of potato tuber, Sol t 1, in adults and children allergic to natural rubber latex.
Source : Allergy. 55(3):266-73, 2000 Mar.

18. Unique Identifier 10072339 ,Status MEDLINE
Authors : Jeannet-Peter N. Piletta-Zanin PA. Hauser C.
Institution :
Division of Immunology and Allergy, Department of Medicine, Hopital Universitaire de Geneve, Geneva, Switzerland.
Title :
Facial dermatitis, contact urticaria, rhinoconjunctivitis, and asthma induced by potato.
Source : American Journal of Contact Dermatitis. 10(1):40-2, 1999 Mar.

19. Unique Identifier 9893201 ,Status MEDLINE
Authors :
Seppala U. Alenius H. Turjanmaa K. Reunala T. Palosuo

T. Kalkkinen N.
Institution : Institute of Biotechnology, University of Helsinki, Finland.
Title :
Identification of patatin as a novel allergen for children with positive skin prick test responses to raw potato.
Source : Journal of Allergy & Clinical Immunology. 103(1 Pt 1):165-71, 1999 Jan.

20. Unique Identifier 9513646, Status MEDLINE
Authors : Zock JP. Hollander A. Heederik D. Douwes J.
Institution :
Environmental and Occupational Health, Wageningen Agricultural University, The Netherlands.

Title :
Acute lung function changes and low endotoxin exposures in the potato processing industry.
Source :
American Journal of Industrial Medicine. 33(4):384-91, 1998 Apr.

21. Unique Identifier 9376085, Status MEDLINE
Authors :
Sabbah A. Sainte-Laudy J. Drouet M. Lauret MG. Loiry ML. el Founini M. Oreac J. Guitton J. Doucet M.
Institution :
Laboratoire d'Immuno-Allergologie-CHRU, Angers.
Title :
[Immuno-biological diagnosis of food allergy]. [Review] [14 refs] [French]
Source : Allergie et Immunologie. 29 Spec No:6-10, 1997

REFERENCES

Jul.

22. Unique Identifier 8839045, Status MEDLINE
Authors : Gaffield W. Keeler RF.
Institution :
Western Regional Research Center, Agricultural Research Service, U.S. Department of Agriculture, Albany, California 94710, USA.
Title :
Induction of terata in hamsters by solanidane alkaloids derived from Solanum tuberosum.
Source :Chemical Research in Toxicology. 9(2):426-33, 1996 Mar.

23. Unique Identifier 8735866 ,Status MEDLINE
Authors :
Zock JP. Doekes G. Heederik D. Van Zuylen M. Wielaard P.
Institution :
Department of Air Quality, Wageningen Agricultural University, The Netherlands.
Title :
Airborne dust antigen exposure and specific IgG response in the potato processing industry.
Source : Clinical & Experimental Allergy. 26(5):542-8, 1996 May.

24. Unique Identifier 8655092, Status MEDLINE
Authors :
Phillips BJ. Hughes JA. Phillips JC. Walters DG. Anderson D. Tahourdin CS.
Institution :BIBRA International, Carshalton, Surrey,

London.
Title :
A study of the toxic hazard that might be associated with the consumption of green potato tops.
Source : Food & Chemical Toxicology. 34(5):439-48, 1996 May.

25. Unique Identifier 8846996 ,Status MEDLINE
Authors : Rayburn JR. Friedman M. Bantle JA.
Institution :
US Environmental Protection Agency, Environmental Research Laboratory, Gulf Breeze, FL 32561, USA.
Title :
Synergistic interaction of glycoalkaloids alpha-chaconine and alpha-solanine on developmental toxicity in Xenopus embryos.
Source : Food & Chemical Toxicology. 33(12):1013-9, 1995 Dec.

26. Unique Identifier 7975140 ,Status MEDLINE
Authors : Spoerke D.
Institution : POISINDEX System, Lakewood, CO 80215-1046.
Title : **The mysterious potato.**
Source : Veterinary & Human Toxicology. 36(4):324-6, 1994 Aug.

27. Unique Identifier 8291267 ,Status MEDLINE
Authors :
Groen K. Pereboom-de Fauw DP. Besamusca P. Beekhof PK. Speijers GJ. Derks HJ.
Authors Full Name :

REFERENCES

Groen, K. Pereboom-de Fauw, D P. Besamusca, P. Beekhof, P K. Speijers, G J. Derks, H J.
Institution :
Unit of Biotransformation, Pharmaco- and Toxicokinetics, National Institute of Public Health and Environmental Protection, Bilthoven, The Netherlands.
Title :
Bioavailability and disposition of **3H-solanine** in rat and hamster.
Source : Xenobiotica. 23(9):995-1005, 1993 Sep.

28. Unique Identifier 7504875 ,Status MEDLINE
Authors : Rotkiewicz T. Szarek J. Tarkowian S.
Institution :
Department of Forensic Veterinary Medicine, University of Agriculture and Technology, Olsztyn, Poland.
Title :
Pathogenic effects of Fusarium sulphureum, Fusarium solani Var. coeruleum and dry rot affected potatoes on the internal organs of rats.
Source : Acta Microbiologica Polonica. 42(1):51-7, 1993.

29. Unique Identifier 8344084 ,Status MEDLINE
Authors : Wang XG.
Institution :
Second Teaching Hospital, Bethune University of Medical Sciences, Jilin.
Title :[Teratogenic effect of potato glycoalkaloids]. [Chinese]
Source : Chung-Hua Fu Chan Ko Tsa Chih [Chinese Journal of Obstetrics & Gynecology]. 28(2):73-5, 121-2, 1993 Feb.

30. Unique Identifier 1564109 ,Status : MEDLINE
Authors :
Hellenas KE. Nyman A. Slanina P. Loof L. Gabrielsson J.
Institution :
Department of Plant Husbandry, Swedish University of Agricultural Sciences, Uppsala.
Title :
Determination of potato glycoalkaloids and their aglycone in blood serum by high-performance liquid chromatography. Application to pharmacokinetic studies in humans.
Source : Journal of Chromatography. A. 573(1):69-78, 1992 Jan 3.

31. Unique Identifier 1894220, Status MEDLINE
Authors : Friedman M. Rayburn JR. Bantle JA.
Institution :
Food Safety Research Unit, USDA-ARS Western Regional Research Center, Albany, CA 94710.
Title :
Developmental toxicology of potato alkaloids in the frog embryo teratogenesis assay--Xenopus (FETAX).
Source : Food & Chemical Toxicology. 29(8):537-47, 1991 Aug.

32. Unique Identifier 2288391 ,Status MEDLINE
Authors : Wuthrich B. Stager J. Johansson SG.
Institution :Allergy Unit, University Hospital, Zurich, Switzerland.
Title :Celery allergy associated with birch and mugwort pollinosis.

REFERENCES

Source : Allergy. 45(8):566-71, 1990 Nov.

33. Unique Identifier 1700766 ,Status MEDLINE
Authors : Wahl R. Lau S. Maasch HJ. Wahn U.
Institution :
Allergopharma Joachim Ganzer KG, Research and Development of Allergen Extract Preparations, Reinbek bei Hamburg, FRG.
Title : **IgE-mediated allergic reactions to potatoes.**
Source :International Archives of Allergy & Applied Immunology. 92(2):168-74, 1990.

34. Unique Identifier 2398519 ,Status MEDLINE
Authors : Hornfeldt CS. Collins JE.
Institution : Hennepin Regional Poison Center, Minneapolis, Minnesota.
Title :**Toxicity of nightshade berries (Solanum dulcamara) in mice.**
Source :Journal of Toxicology - Clinical Toxicology. 28(2):185-92, 1990.

35. Unique Identifier 2378105,Status MEDLINE
Authors :
 Tsyganenko OI. Rymar'-Shcherbina NB. Gulich NL. Emchenko NL. Lapchenko VS. Stakhurskaia LV. Nikolenko TN. Stadnichuk N. Mikhaliuk EN. Vashchenko NV. et al.
Title:
[The hygienic evaluation of potato treated with the preparation Polikar made from the byproducts of soda manufacture]. [Russian]

PREVENTING HEADACHES AND MIGRAINES

Source : Voprosy Pitaniia. (2):74-7, 1990 Mar-Apr.

36. Unique Identifier 2341091 ,Status MEDLINE
Authors : Hageman G. Hermans R. ten Hoor F. Kleinjans J.
Institution :
Department of Biological Health Science, University of Limburg, Maastricht, The Netherlands.
Title :
Mutagenicity of deep-frying fat, and evaluation of urine mutagenicity after consumption of fried potatoes.
Source : Food & Chemical Toxicology. 28(2):75-80, 1990 Feb.

37. Unique Identifier 2107539 ,Status MEDLINE
Authors : Matteo A. Sarles H.
Institution :
Clinique des Maladies de l'Appareil Digestif et de la Nutrition, Hopital Sainte Marguerite, Marseille, France.
Title : Is food allergy a cause of acute pancreatitis?.
Source : Pancreas. 5(2):234-7, 1990 Mar.

38. Unique Identifier 2617053 ,Status MEDLINE
Authors : Kucharski R. Marchwinska E. Piesak Z.
Title :
[Possibility of potato growing in the central part of the Katowice province in the aspect of population exposure to harmful substances]. [Polish]
Source : Roczniki Panstwowego Zakladu Higieny. 40(2):131-6, 1989.

REFERENCES

39. Unique Identifier 2692473 ,Status MEDLINE
Authors:
Quirce S. Diez Gomez ML. Hinojosa M. Cuevas M. Urena V. Rivas MF. Puyana J. Cuesta J. Losada E.
Institution :
Servicios de Alergia e Inmunologia, Hospital Ramon y Cajal, Madrid, Spain.
Title : Housewives with raw potato-induced bronchial asthma.
Source : Allergy. 44(8):532-6, 1989 Nov.

40. Unique Identifier 3071288 ,Status MEDLINE
Authors : Laurent J. Wierzbicki N. Rostoker G. Lang P. Lagrue G.
Institution :
Service de Nephrologie, Hopital Henri-Mondor, Creteil.
Title :
[Idiopathic nephrotic syndrome and food hypersensitivity. Value of an exclusion diet]. [Review] [17 refs] [French]
Source : Archives Francaises de Pediatrie. 45(10):815-9, 1988 Dec.

41. Unique Identifier 3206859 ,Status MEDLINE
Authors : Gritsevskaia IL.
Title :
[Toxicological and hygienic evaluation of potatoes grown with the use of a pesticide croneton]. [Russian]
Source : Voprosy Pitaniia. (3):58-61, 1988 May-Jun.

42. Unique Identifier 3612898 ,Status MEDLINE
Authors : Baker D. Keeler R. Gaffield W.
Title :

PREVENTING HEADACHES AND MIGRAINES

Lesions of potato sprout and extracted potato sprout alkaloid toxicity in Syrian hamsters.
Source :
Journal of Toxicology - Clinical Toxicology. 25(3):199-208, 1987.

43. Unique Identifier 2866606 ,Status MEDLINE
Authors : Hemminki K. Vineis P.
Title :
Extrapolation of the evidence on **teratogenicity of chemicals between humans and experimental animals:** chemicals other than drugs. [Review] [60 refs]
Source :Teratogenesis, Carcinogenesis, & Mutagenesis. 5(4):251-318, 1985.

44. Unique Identifier 6515563 ,Status MEDLINE
Authors : Renwick JH. Claringbold WD. Earthy ME. Few JD. McLean AC.
Title :
Neural-tube defects produced in Syrian hamsters by potato glycoalkaloids.
Source : Teratology. 30(3):371-81, 1984 Dec.

45. Unique Identifier 6338654 ,Status MEDLINE
Authors : Dalvi RR. Bowie WC.
Title : **Toxicology of solanine: an overview.** [Review] [23 refs]
Source :
Veterinary & Human Toxicology. 25(1):13-5, 1983 Feb.

46. Unique Identifier 7423541 ,Status MEDLINE
Authors : Bergers WW. Alink GM.

REFERENCES

Title :
Toxic effect of the glycoalkaloids solanine and tomatine on cultured neonatal rat heart cells.
Source : Toxicology Letters. 6(1):29-32, 1980 Jun.

47. Unique Identifier 1138407 ,Status MEDLINE
Authors : Mun AM. Barden ES. Wilson JM. Hogan JM.
Title :
Teratogenic effects in early chick embryos of solanine and glycoalkaloids from potatoes infected with late-blight, Phytophthora infestans.
Source : Teratology. 11(1):73-8, 1975 Feb.

Studies on **Bananas**

1.Unique Identifier
 21732181
Status
 MEDLINE
Authors
  Menezes EW. Tadini CC. Tribess TB. Zuleta A. Binaghi J. Pak N. Vera G. Dan MC. Bertolini AC. Cordenunsi BR. Lajolo FM.
Authors Full Name
  Menezes, Elizabete Wenzel. Tadini, Carmen Cecilia. Tribess, Tatiana Beatris. Zuleta, Angela. Binaghi, Julieta. Pak, Nelly. Vera, Gloria. Dan, Milana Cara Tanasov. Bertolini, Andrea C. Cordenunsi, Beatriz Rosana. Lajolo, Franco M.
Institution

PREVENTING HEADACHES AND MIGRAINES

Department of Food and Experimental Nutrition, Faculty of Pharmaceutical Sciences, University of Sao Paulo, Sao Paulo, Brazil. wenzelde@usp.br
Title
Chemical composition and nutritional value of unripe banana flour (Musa acuminata, var. Nanicao).
Source
Plant Foods for Human Nutrition. 66(3):231-7, 2011 Sep.

2.Unique Identifier
21915417
Status
MEDLINE
Authors
Saravanan K. Aradhya SM.
Authors Full Name
Saravanan, Kandasamy. Aradhya, Somaradhya Mallikarjuna.
Institution
Fruit and Vegetable Technology Department, Central Food Technological Research Institute, a constituent laboratory of the Council of Scientific and Industrial Research, Mysore, 570 020, Karnataka, India.
Title
Potential nutraceutical food beverage with antioxidant properties from banana plant bio-waste (pseudostem and rhizome).
Source
Food & Function. 2(10):603-10, 2011 Oct.

REFERENCES

3. Unique Identifier
   21445854
Status
   MEDLINE
Authors
   Shiga TM. Soares CA. Nascimento JR. Purgatto E. Lajolo FM. Cordenunsi BR.
Authors Full Name
   Shiga, Tania M. Soares, Claudineia A. Nascimento, Joao Ro. Purgatto, Eduardo. Lajolo, Franco M. Cordenunsi, Beatriz R.
Institution
   Laboratorio de Quimica, Bioquimica e Biologia Molecular de Alimentos, Departamento de Alimentos e Nutricao Experimental, FCF, Universidade de Sao Paulo, Avenida Professor Lineu Prestes 580, Bloco 14, CEP 05508-000, Sao Paulo, SP, Brazil.

4. Unique Identifier
   21384381
Status
   MEDLINE
Authors
   Bhaskar JJ. Salimath PV. Nandini CD.
Authors Full Name
   Bhaskar, Jamuna J. Salimath, Paramahans V. Nandini, Chilkunda D.
Institution

PREVENTING HEADACHES AND MIGRAINES

Department of Biochemistry and Nutrition, Central Food Technological Research Institute, Mysore 570020, India.

5. Unique Identifier
   21369778
   Status
   MEDLINE
   Authors
   Sundaram S. Anjum S. Dwivedi P. Rai GK.
   Authors Full Name
   Sundaram, Shanthy. Anjum, Shadma. Dwivedi, Priyanka. Rai, Gyanendra Kumar.
   Institution
   Centre for Biotechnology, University of Allahabad, Allahabad 211002, India. shanthy_s@rediffmail.com

6. Unique Identifier
   21598798
   Status
   MEDLINE
   Authors
   Jadhav UU. Dawkar VV. Jadhav MU. Govindwar SP.
   Authors Full Name
   Jadhav, Umesh U. Dawkar, Vishal V. Jadhav, Mital U. Govindwar, Sanjay P.
   Institution
   Department of Biochemistry, Shivaji University, Kolhapur, India.
   Title
   Decolorization of the textile dyes using purified banana pulp polyphenol oxidase.

REFERENCES

Source
  International Journal of Phytoremediation. 13(4):357-72, 2011 Apr.

7. Unique Identifier
   21405133
Status
   MEDLINE
Authors
   Saravanan K. Aradhya SM.
Authors Full Name
   Saravanan, K. Aradhya, S M.
Institution
   Fruit and Vegetable Technology Department, Central Food Technological Research Institute, Mysore, Karnataka, India.

Coeliac Disease books
1. The Everyday Wheat-Free and Gluten-Free Cookbook, by MichelleBerriedale-Johnson.2003. ISBN 1 898697 90 6.

2. 200 Gluten-Free Recipes, by Louise Blair, 2011. ISBN 13:978-0-600-62268-0.

3. Your Guide to Coeliac Disease by Professor Peter Howdle, 2007. ISBN -13: 9-780430-928851. Royal Society of Medicine.

PREVENTING HEADACHES AND MIGRAINES

www.ingramcontent.com/pod-product-compliance
Lightning Source LLC
Chambersburg PA
CBHW022013170526
45157CB00003B/1225